New Insights Into Protected Area Management and Conservation Biology

Edited by Levente Hufnagel

Published in London, United Kingdom

New Insights Into Protected Area Management and Conservation Biology
http://dx.doi.org/10.5772/intechopen.104052
Edited by Levente Hufnagel

Contributors
Chandima Fernando, Dale W. Sparks, Duminda S.B. Dissanayake, Franco P. Mbise, Helle Mäemets, Javier Ferreiro da Costa, Levente Hufnagel, Makuach Robert Athian, Pablo Ramil-Rego, Pasquale Tiberio Moilinga, Scott Kennedy, Virgil Brack

First published in London, United Kingdom, 2024 by IntechOpen
IntechOpen is the global imprint of INTECHOPEN LIMITED, registered in England and Wales, registration number: 11086078, 5 Princes Gate Court, London, SW7 2QJ, United Kingdom

British Library Cataloguing-in-Publication Data
A catalogue record for this book is available from the British Library

Additional hard and PDF copies can be obtained from orders@intechopen.com

New Insights Into Protected Area Management and Conservation Biology
Edited by Levente Hufnagel
p. cm.
Print ISBN 978-1-83768-354-3
Online ISBN 978-1-83768-355-0
eBook (PDF) ISBN 978-1-83768-356-7

Meet the editor

Levente Hufnagel is a full professor and the head of the Research Institute of Multidisciplinary Ecotheology, John Wesley Theological College, Budapest, Hungary, working on ecology, biogeography, ecological research methodology, ecotheology, and sustainability. He has more than 20 years of experience leading Hungarian academic institutions and teaching and supervising Ph.D., MSc, and BSc students from various social and cultural backgrounds. He has more than 260 scientific publications (in both aquatic and terrestrial ecological aspects of plants, animals and microbes, at community as well as population level) to his credit. As a participant in several big ecological research and development projects, Dr. Hufnagel has significant experience in multidisciplinary cooperation. He has collaborated with more than 200 coauthors and is editor-in-chief of an international scientific journal. Dr. Hufnagel graduated from Eötvös Lorand University, Budapest, Hungary, with a master's degree in Ecology and Evolutionary Biology and a Ph.D. in Hydrobiology. He obtained an additional Ph.D. in Agricultural Science from Szent István University, Gödöllő, Hungary, and several other degrees from Corvinus University of Budapest and Adventist Theological College, Hungary.

Contents

Preface

Humankind is an important but only a small part of the biosphere; the human race is just one of the many millions of species living on our planet. The existence of our species and the functionality and sustainability of our society, economy, and culture depend on the ecosystem services provided by the rest of the biosphere.

The biosphere's state of health and its ability to provide services depend on the extent of the living communities of natural and near-natural habitats, their biomass, productivity, biological activity, and the high biodiversity that ensures their reliability and flexibility.

The process of the historical unfolding of the global ecological crisis, at different stages, necessitated nature conservation activities of different levels and types (species, habitat, and biodiversity protection, in situ and ex-situ approaches, and passive or active management tools).

In the maintenance and conservation of our planet's biodiversity and ecosystem services, knowledge of current ecological and biogeographical patterns, changes in flora, fauna, and human society have outstanding importance.

This book, "New Insights into Protected Area Management and Conservation Biology", gives a useful report about this important applied scientific area. As such, it is sure to be a valuable resource for readers looking to gain some insight into the recent problems of conservation efforts.

Levente Hufnagel
Research Institute of Multidisciplinary Ecotheology,
John Wesley Theological College,
Budapest, Hungary

Section 1

Protected Area Management

Chapter 1

Introductory Chapter: Biodiversity Conservation Levels and Approaches in Today's Global Ecological Crisis

Levente Hufnagel

1. Introduction

Human society is part of the biosphere, the human race is just one of the many millions of species in the biosphere. The existence of humanity, and the functionality and sustainability of its society and economy depend on the ecosystem services provided by the rest of the biosphere [1]. These services ensure breathable air, self-purification of waters and soils, a climate system suitable for life, and food, but a significant part of the energy carriers and building materials are also the product of the past operation of the biosphere.

The biosphere's state of health and its ability to provide services depend on the extent of the living communities of natural and near-natural habitats, their biomass, productivity, biological activity, and the high biodiversity that ensures their reliability and flexibility.

The global overpopulation of the human race has resulted in the shrinking of natural habitats, changes in land use and vegetation coverage, urbanization, and the growth of low-diversity agricultural areas, which on the one hand has led to a decrease in ecosystem services, a mass extinction wave of species, and on the other hand to additional problems resulting in a global ecological crisis, and climate change, which led to social and public health problems, socioeconomic dangers, and an increase in risks [2].

The transition to a sustainable society means the way out of the global ecological crisis [3, 4]. However, the sustainable society is currently not a real way of functioning, not an existing economic-social system, not a well-structured system of existing technologies and methods, but unfortunately only a slogan.

We know that in order to realize a sustainable global society, we need to achieve changes at several different organizational levels at the same time, and these are as follows:

- sustainable lifestyle of individuals and families

- the sustainable operation of institutions, companies, social organizations, churches, local governments, and their activities supporting a sustainable lifestyle of families.

- local-, regional-, continental-, and global-level legislation and law enforcement, concluding international conventions and establishing and strengthening international organizations that implement them, ensuring the legal and coordination conditions of a sustainable society.

We know that efforts to solve global problems during the transition to a sustainable information society must cover the following areas:

- Protecting the natural environment and habitats
- Maintaining the functionality and biodiversity of the biosphere
- Protection of the climate system and air quality
- Energy supply, and transition to renewables
- Water supply and protection of water resources
- Food supply and food safety
- Waste management and recycling
- Eliminate terrorism, organized crime, corrupt political systems, and war
- Fighting diseases (epidemics, environmental damage, civilizational damage)
- Eliminating poverty, reducing social differences, tensions, and aggression
- Ensuring social equality of opportunity and legal equality
- Modernization of education
- Boosting scientific research, operating knowledge transfer channels
- Ensuring democracy and operating its institutional system

Global problems can only be tackled through globally coordinated efforts. The institutional system performing macro-level coordination tasks for the creation of a future sustainable society must fulfill the following (today not yet fulfilled worldwide) functions:

- Operation of the joint legislation and expert decision support system.
- Ensuring global publicity of the comprehensive information necessary for decisions.
- Operation of the global democratic institution system of decision-making.
- Ensuring the implementation of the decisions made.

- Ensuring control and justice related to all of this.
- Enforcing the common global interests of humanity against local and group interests.
- Representing and embodying the unity of humanity as a whole.

The common goals to be achieved are obvious:

- Preservation and improvement of the biosphere's effective self-regulation capacity and biodiversity.
- Creating, organizing, and operating a sustainable society.
- Promotion of scientific knowledge and technical development.
- The cultural, spiritual, and moral upliftment of humanity.
- Increasing people's level of satisfaction (feeling of happiness).
- Possible success indicators can also be identified:
- The total biomass of the biosphere and its biological activity
- Biodiversity of the biosphere
- Self-regulating capacity, long-term sustainability, adaptability, flexibility, and defense capacity of the biosphere and human society as part of it.
- The ability of human society to create, develop, and preserve culture (science, art, knowledge transfer).
- Expected average lifespan of people at birth (physical and mental health, healthy environment).
- People's average level of education, general education, and awareness.
- The possibility of developing individual talents.
- The average standard of living, quality of life.

2. The special tasks and response levels of biodiversity conservation

The process of the historical unfolding of the global ecological crisis, at different stages, necessitated nature conservation activities of different levels and types. As the problems worsen, the protection activities must also develop and adapt to new challenges. In this historical adaptation process, seven levels or eras can be identified. However, the newer approaches can be successful not instead of the older ones, but only in addition to them and based on them.

2.1 Passive *in situ* species protection

The reduction of habitats and the increase of human endangering activities first manifested themselves in the conspicuous decline of large, mainly predatory animals and old large trees. It is clear that the first reactions of nature conservation also aim to protect these endangered and rare species, primarily by limiting and prohibiting their deliberate and direct destruction (hunting, cutting, collecting, trading) in their natural habitat. This level of protection is still important and essential in case of large marine and terrestrial mammals, birds, reptiles, amphibians, and rare plants as well.

2.2 Active *in situ* species protection

It soon became obvious that passive *in situ* species protection is not sufficient to deal with the problems in many cases, because the destruction of species and the reduction of their populations are not only caused by their direct destruction, but rather by the reduction of their life possibilities and their survival in critical periods. Their active protection in their own habitat can be served more efficiently by providing winter feeding places, summer watering holes, artificial nesting opportunities, and migration routes.

2.3 Active *ex situ* species protection

Unfortunately, it happened in more and more cases that both passive and active methods of *in situ* species protection failed, and the survival of some endangered species could no longer be ensured due to the disappearance or shrinking of their natural habitat. In such cases, as a last desperate attempt, the means of active *ex situ* species protection had to be and should be used, and the last specimens of endangered species had to be propagated under artificial conditions, in zoos, botanic gardens, preserved and maintained in special reserves, gene banks, and then reintroduced after the restoration of their habitat.

2.4 Passive *in situ* habitat protection

Already in the earliest times, it became obvious that the protection of species is most easily and effectively possible in their original natural habitat, together with their living community. This, in turn, means that the protection of non-endangered organisms and the protection of the association and ecosystem as a whole must be ensured in the habitat of protected organisms. The first national parks, nature reserves, biosphere reserves, and habitat protection programs were launched under this concept, but again, for the first time, only protection against direct human damage (logging, hunting) was implemented, and in many cases only partially.

2.5 Active *in situ* habitat protection

In large areas, a diverse mosaic of ecosystems in their natural state is capable of self-sustaining operation if the harmful human activity is excluded. In many cases, however, there is only the possibility of preserving a small piece of nature in smaller patches of habitat surrounded by human settlements and agricultural areas, where the natural self-sustaining mechanisms have already been damaged. The active management of such nature conservation areas is essential, as invasive species must be actively kept away, missing predators must be actively replaced by herbivore population control, the missed visits of grazing animals must be compensated for by mowing, forest fires must be curbed, or need to be maintained the dynamics of special communities by artificial controlled fires, water replacement may be necessary, etc.

2.6 Extended biodiversity protection

Nowadays, it has become clear that no matter how much we try to increase the number and extent of national parks, and to improve the quality of the professional work carried out in them, these efforts alone cannot effectively stop the decrease in biodiversity caused by human overpopulation and environmental pollution. Protection activities must also be extended to artificial ecosystems in the immediate vicinity of national parks and protected areas, by establishing buffer zones around protected areas, where agricultural activity should only be allowed in the form of chemical-free organic farming, the spread of urbanization must be stopped, and the establishment of mines and industrial areas must be limited, and the protection of biodiversity must also be ensured in settlement areas (by establishing parks, artificial lakes, green roofs, tree rows, lawn areas, and methods of active species protection). Extended biodiversity protection must cover the old varieties of cultivated plants and farm animals, traditional farming methods, and the traditional ecological knowledge of natural peoples, as well as the preservation of the spiritual and material memories of our cultural heritage. Efforts must be made to ensure that the culture and traditional lifestyle of natural peoples, which is in line with the approach of a sustainable society, can be preserved in its original form as our common treasure. All of these require a change of attitude in the field of tourism as well, and trophy hunting tourism can for example be replaced by ecotourism (as well as wildlife management for the purpose of nature conservation), which can be an important tool for spreading knowledge about nature and ethnography, and can be connected to nature photography, forest schools, etc. In the course of realizing a sustainable society, we must actually introduce the transition to environmentally friendly, sustainable farming not only in the buffer zones, but in the entire area of our planet, and the need for the general expansion of various methods of biodiversity protection.

2.7 Active *ex situ* habitat protection

The phenomenon of global climate change draws our attention to the fact that some long-term effects of human activity and natural processes make the work in national parks and their buffer zones and the general extension of biodiversity protection insufficient. Global climate change changes the boundaries of the climate zones suitable for individual ecosystems and communities [5, 6], shifts them in a polar or east-west direction, pushes the vertical zonation of mountains upwards, changes oceanic and continental effects, results in the transformation of local weather and groundwater conditions, the water flow of rivers, the water balance of lakes, the location and water level of seashores.

Instead of the currently prevailing "*in situ* conservation", that is, the preservation of existing ecological conditions in current habitats, the goal of nature conservation can only be to preserve the functionality, self-regulating capacity and biological diversity of the biosphere. This can be achieved by preventing harmful anthropogenic effects and actively helping the natural adaptation processes of ecological systems (migration, migration routes, area change). In order to solve this problem, active nature conservation work in the nature of "eco-engineering" cannot be avoided. This work can basically be divided into two sub-tasks:

- promoting the creation of natural and close-to-natural communities of organisms that are suitable for (and adapted to) changing (i.e., future) climate of the given conservation area,

- providing an escape route for communities of organisms that currently and in the past live in the given area, but are unable to adapt to its changing ecological conditions due to climate change, if necessary through conscious plantings and the establishment of eco-corridors.

3. Financial issues

The broadening and deepening of active nature conservation interventions and scientific research work also raise financial issues. In this connection, the apparent interests of different human activities and sectors may conflict, which make it necessary to express the effects and consequences in monetary terms. The tools of ecological economics provide an opportunity for this. The monetary value of biosphere services per unit of time and area can be estimated or approximated using several approaches. From this point of view, the article by Costanza et al. published in Nature in 1997 should be highlighted, and since then there have been many further developers and critics [7]. Opponents of monetary valuation of natural capital and biosphere services are mostly afraid that this method of valuation suggests that the biosphere can be bought or its functions can be replaced with money. Of course, it is not about that, it's just about using money as a general measure of value and "unit of measurement" to manage different habitat types and the importance and effects of their services in a way that is suitable for comparison.

Market-friendly solutions can be introduced based on the example of carbon dioxide emissions trading, which could play an important role in climate change mitigation and adaptation. Forest plantations for climate protection purposes, translocation projects for nature conservation purposes, the withdrawal of land from intensive cultivation, or the development of agriculture in accordance with ecological conditions, and the spread of ecotourism can be boosted with the help of these tools.

Author details

Levente Hufnagel
Research Institute of Multidisciplinary Ecotheology, John Wesley Theological College, Budapest, Hungary

*Address all correspondence to: leventehufnagel@gmail.com

References

[1] Palliwoda J, Fischer J, Felipe-Lucia MR, Palomo I, Neugarten R, Büermann A, et al. Ecosystem service coproduction across the zones of biosphere reserves in Europe. Ecosystems and People. 2021;**17**(1):491-506. DOI: 10.1080/26395916.2021.1968501

[2] Cafaro P, Hansson P, Götmark F. Overpopulation is a major cause of biodiversity loss and smaller human populations are necessary to preserve what is left. Biological Conservation. 2022;**272**:109646. ISSN 0006-3207. DOI: 10.1016/j.biocon.2022.109646

[3] Hufnagel L, Pálinkás M, Mics F, Homoródi R. Introductory chapter: The present global ecological crisis in the light of the mass extinctions of earth history. In: Hufnagel L, editor. Changing Ecosystems and Their Services. London, UK: IntechOpen; 2020. p. 141. Paper: Chapter 1

[4] Bastante-Ceca J, María Fuentes-Bargues L, José Florin-Constantin M, Latu C, et al. Introductory chapter: The need to change the paradigm - sustainability and development at the 21st century. In: María JB-C, Jose LF-B, Levente H, Florin-Constantin M, Corneliu L, editors. Sustainability Assessment at the 21st Century. London, UK: IntechOpen; 2020. p. 185. Paper: Chapter 1

[5] Garamvölgyi Á, Hufnagel L. Impacts of climate change on vegetation distribution no. 1 climate change induced vegetation shifts in the Palearctic region. Applied Ecology and Environmental Research. 2013;**11**(1):79-122

[6] Hufnagel L, Garamvölgyi Á. Impacts of climate change on vegetation distribution no. 2 - climate change induced vegetation shifts in the new world. Applied Ecology and Environmental Research. 2014;**12**(2):355-422

[7] Costanza R, d'Arge R, de Groot R, Farber S, Grasso M, Hannon B, et al. The value of the world's ecosystem services and natural capital. Nature. 1997;**387**:253-260

Chapter 2

Protecting Wetlands: Insights from the Northern Iberian Peninsula (Galicia, NW Spain)

Javier Ferreiro da Costa and Pablo Ramil-Rego

Abstract

Wetlands are a key tool for environment conservation policy. They harbour important biodiversity values such as priority habitats and fragile species, reduce the impacts of floods, improve water quality, absorb pollutants, and protect shores from climate change effects, also acting as carbon reservoirs in the medium and long term. From an international point of view, those sites containing representative, rare or unique wetlands, are designated under Ramsar Convention, which was signed in 1971, being added to the Convention's List of Wetlands of International Importance and become known as Ramsar sites. More than 50 years after the signing of Ramsar Convention, its degree of application is very uneven across the different territories. This paper analyses the situation from the Atlantic area of the Iberian Peninsula, and specifically from Galicia, a territory that has a large number of wetlands, both terrestrial, marine, underground and artificial, with sites of high value for biodiversity and natural heritage conservation, but where there is no adequate protection over them, documented by the presence of a large number of anthropic impacts that is leading to biodiversity deterioration, habitat destruction and species decline.

Keywords: wetlands, conservation, management, protection, Ramsar, biodiversity

1. Introduction

During the last decades of the 20th century, a paradigm shift in the social assessment of wetlands happened, as key components of biodiversity and donors of ecosystem services, leading to the implementation of various mechanisms, instruments and agreements for their conservation. The most relevant example was the signing of an International Convention in 1971 in the Iranian city of Ramsar, for the conservation of wetlands of international importance as waterfowl habitats (known as 'Ramsar Convention'), which made wetlands the only large ecosystem subject to an international agreement aimed at promoting its protection and rational use [1].

The signing of the Ramsar Convention would serve as an impulse for the development of new initiatives from an international perspective, as well as at a national, regional, or local level. In this way the International Union for the Conservation of Nature (IUCN) applied, considering the conservation of wetlands

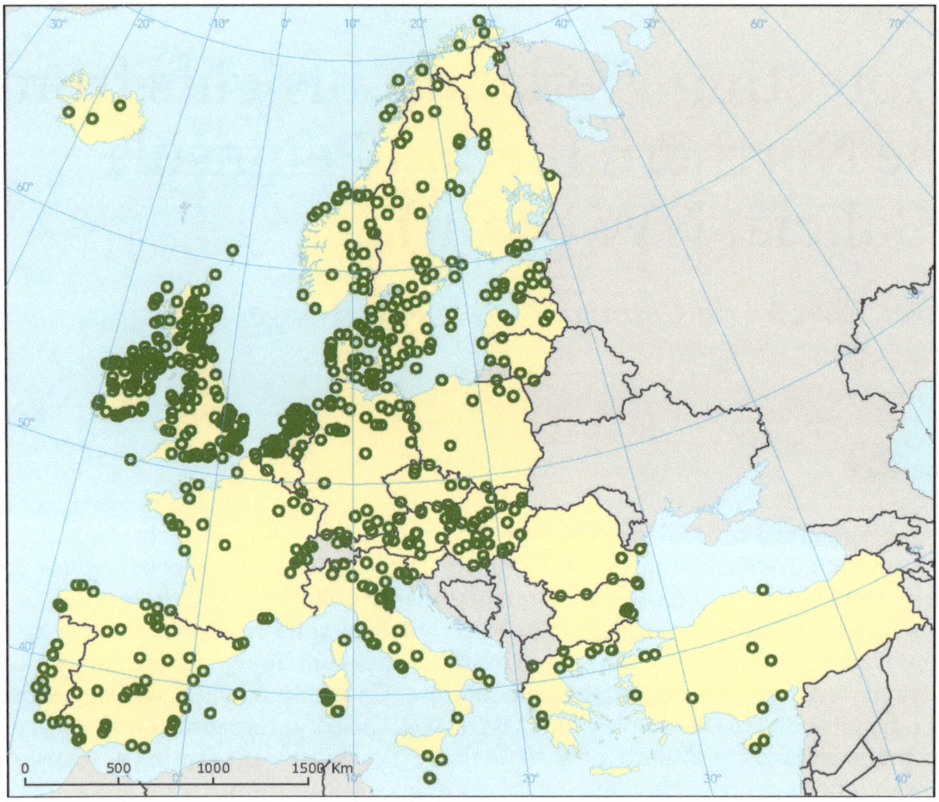

Figure 1.
Distribution of Ramsar sites within EU Member states [2].

of vital importance, since they contribute to the conservation of biological diversity and cultural heritage, in addition to having an important role for human beings in terms of climate regulation, carbon sink, recharging aquifers, flood mitigation and other processes, as well as providing important provision services as they are an important source of directly exploitable resources. From the signing of the Ramsar Convention, wetlands would begin to be integrated into the legal framework of different countries. In 1990, after 15 years of the entry into force of the Ramsar Convention, the IV Conference of the Contracting Parties (COP4) of the Ramsar Convention was held in Montreux (Switzerland), bringing together 56 of the 59 signatory countries of the convention, approving the designation of 53 new wetlands of International Importance, which added to those declared since 1974, reached 533 designations (**Figure 1**).

In 1992, the United Nations Conference on Environment and Development (UNCED), also known as the 'Earth Summit', was held in Rio de Janeiro. At the Earth Summit, 5 highly relevant documents were approved that marked environmental policy in the following decades: Rio Declaration on the environment and development (which established the concept of sustainable development), Agenda 21 (which listed 2,500 recommendations relating to the implementation of the declaration's principles), the United Nations Framework Convention on Climate Change (which led to the signing in 1997 of the Kyoto Protocol), the Declaration of Principles on Forests, and finally the Convention on Biological Diversity (CBD).

The European Union played a relevant role in the Rio Earth Summit, and at the same time promoted new environmental regulations. Thus, the Birds Directive (Directive 79/409/EEC), aimed at the conservation of wild birds, was complemented by the promulgation of the Habitats Directive (Directive 92/43/EEC), relating to the conservation of natural habitats and wild fauna and flora, in which Natura 2000 was created. Subsequently, on 05/29/1995 the European Commission approved a communication from the Commission to the Council and the European Parliament "Wise use and conservation of wetlands (COM(95) 189 final)", where it was assumed that wetlands represent one of the most important, most threatened and most common habitats in all the countries of the European Union. All this European legal framework granted the consideration of habitats of community interest to the natural and semi-natural wetland habitats established in the Ramsar classification. The protection and conservation of wetlands within the framework of the European Union was completed through a third regulation, the Water Framework Directive (Directive 2000/60/EC).

In Spain, the new orientations of environmental policy regarding wetlands were embodied by the very first time in Law 4/1989, which in its article 25 it contemplated the creation of the Spanish Wetland Inventory (SWI), whose preparation corresponded to the Spanish regions. This new framework was continued by the Spanish Strategy for the Conservation and Sustainable Use of Biological Diversity [3], which established at the end of the 20th century the general framework of Spanish conservation policy, as an application of the Convention on Biological Diversity (Rio de Janeiro, 1992). The application of this Strategy was proposed through the development of Sectoral Action Plans, which in the framework of aquatic ecosystems led to the preparation of the Spanish Strategic Plan for the Conservation and Rational Use of Wetlands [4]. This plan was inspired by the First Strategic Plan of the Ramsar Convention 1997–2002 [5], as well as the postulates of the IUCN, thus contributing to compliance with the international commitments acquired by the Spanish government. However, the progress of the SWI set out in Law 4/1989 followed a slow development, as its regulation was not established until 15 years later, through Royal Decree 435/2004. Finally, Law 42/2007 came to recognize the Wetlands of the Ramsar Convention within the Spanish legislative framework as Areas Protected by International Instruments (APII), establishing additional protection measures for Spanish wetlands, mainly those included within one of the European or Spanish protection categories that were also established in the Law.

In application of Law 4/1989, the different Spanish regions started to take their first steps in terms of wetland protection. In Galicia, the regional legal framework in terms of protecting wetlands would take an important step with the promulgation of Law 9/2001, since it included a total of 9 categories of Protected Natural Areas (PNA), among which was the specific category of Protected Wetlands, setting an unparalleled example in Spain at that time. All these regulatory changes carried out in Galicia since the end of the 1980s were accompanied by different actions regarding the characterization and monitoring of waterfowl and the ecological restoration of wetlands [6], which showed a remarkable change of direction compared to the destructive policies on wetlands, developed under Franco's dictatorship and continued during the first stages of Spanish democracy.

In this paper, as one of the first planning and dissemination tasks developed by LIFE INSULAR project (LIFE20 NAT/ES/001007), we analyse the management, methods and processes on Protected Wetlands biodiversity in Galicia, assessing the future challenges and perspectives that arise for this protected area category.

The document assesses the progress in biodiversity conservation and management that has been made on Protected Wetlands over time, including the inventorying of wetlands, as well as their legal provisions and designations, the trends and changes of their limits because of several reasons, their improper use and management measures, as well as the consequences of all of these aspects which results in a loss of quality in the legal framework of Protected Wetlands.

2. Study area

Galicia is a Spanish region that occupies an area of 3 million hectares and is located at the NW end of the Iberian Peninsula (**Figure 2**). Its Nomenclature of Territorial Units for Statistics code is ES11. It is a territory of strong biogeographical and landscape contrasts [7–10], mostly included in the Atlantic biogeographical region, but with a significant part of its territory in the Mediterranean region. It is bounded by the Cantabrian Sea on the North, by the Atlantic Ocean on the West, by the Spanish regions of Asturias and Castilla y León on the East, and by the Portuguese region of Regiao Norte on the South. The marine waters surround Galician extensive northern and western coastline, while eastern edge encompasses the western end of the Cantabrian Mountains, which are diluted across the different Galician mountain ranges that separate the Littoral Galicia from the Interior Galicia. The first one is configured by a succession of short river valleys separated from each other by small reliefs. The second one is represented by large sedimentary basins, which mean a landscape with large horizontal extensions, drained by the tributaries of Miño, Sil and Limia rivers. In some sections these pass throughout canyons, especially the main streams of Miño and Sil rivers.

This huge heterogeneity has determined a high level of biodiversity in the Galician territory [11–21], both from the point of view of the diversity of wild species present (an especially the protected and catalogued species), as well as from the wide range of habitat and ecosystem types.

Figure 2.
Geographic framework of Galicia in NW Iberian Peninsula. Created by the authors.

3. The importance of wetland inventories

The Ramsar Convention recognized from the beginning the importance of national wetland inventories as essential instruments for shaping policies and other measures aimed at achieving the conservation and wise use of wetlands. Already at the first meeting of the Conference of the Contracting Parties (COP1, Cagliari, 1980), the Parties were convinced that national wetland policies should be based on a nation-wide inventory of wetlands and their resources (Recommendation 1.5). This recognition of the value of national wetland inventories has been periodically reiterated at subsequent COPs, inter alia in the Annex to Recommendation 2.3 (COP2, Groningen, 1984), Recommendation 4.6 (COP4, Montreux, 1990), Resolution 5.3 (COP5, Kushiro, 1993) and Resolution VI.12 (COP6, Brisbane, 1996). So national wetland inventories, in addition to being an essential basis for the formulation of national wetland policies, are also considered important, among other things, to detect sites that can be included in the List of Wetlands of International Importance (the Ramsar List); to quantify the world's wetland resources, in order to assess their status and patterns; to determine which wetlands need to be restored and to carry out risk and vulnerability assessments [22].

The first studies for the cataloguing of the Iberian wetlands were developed in the mid-20th century [23]. After the signing of the Ramsar Convention, the drafting of different studies was promoted, resulting in the first different provincial inventories of wetlands and some synthesis works [24–28]. Finally, the importance of carrying out an inventory of wetlands was included in the Spanish Strategic Plan for the Conservation and Rational Use of Wetlands [4], developing its technical aspects through Royal Decree 435/2004, regulating the SWI, which is based on the regional inventories and consequently on the regional lists of wetlands that are prepared from the available data. The number of Spanish regions that have carried out a proper inventory of their wetlands is very small, and in most cases these inventories have not been updated and published. So, the wetlands included in the SWI are limited, in most of the regions, to the Ramsar list, a fact that limits the effectiveness of the inventory to comply with the aspects set out in the Spanish regulations and in the agreements derived from the Ramsar Convention.

Although there is not a periodically updated list of wetlands in Spain, compilations made in 2003 [29] indicate that Spanish wetlands should exceed 5,000 sites, of which more than 2,027 sites would be located in the 7 regions of Northern Spain (**Figure 3**), which are Galicia, Asturias, Cantabria, Castilla y León, Euskadi, Navarra and La Rioja. No further works have been made estimating the Spanish wetlands, so nowadays there is no updated information about this. Moreover, the number of wetlands in Spain has not been adequately collected in the official inventories managed by the Spanish and regional governments. Thus, the SWI shows a very small number of wetlands in the Northern Iberian regions, as in many cases only includes wetlands that are part of the Spanish Ramsar List.

Both in the SWI and in the List of Ramsar Wetlands of Spain, the mountains of the Atlantic-Cantabrian area show a meagre representation, which in no case reflects the high diversity and richness that these territories possess [15–17, 20, 21, 30–35], where important representations of wetland types of high environmental singularity at an international level, such as peatlands that include different types of ecosystems (blanket bogs, raised bogs, fens, floating mires, bog woodlands, etc.), as well as

Figure 3.
Number of wetlands identified in each Spanish region during the period 2001–2003, coinciding with the carrying out of GWI [29].

different types of scrublands, wet heaths and hygrophilous forests. The deficient content of the SWI reflects the inefficient inventorying and management of the biodiversity components which also has a negative impact on its conservation, even more so when these territories can be immersed in the evaluation of the effects that could derive from the implementation of new projects for power generation or from the promotion of intensive forestry and livestock farms [36, 37].

During the two last decades of the 20th century, an important work aimed at the legal protection and rational use of wetlands began in Northern Spain. A large number of wetlands were included in the first Natura 2000 proposals, which between 1997 and 2004 formed almost the entirety of its current delimitation. This contrasted with those designed within the framework of the SWI, where the number of wetlands registered to date is very small, leaving out wetlands that have environmental protection status derived from their inclusion in different figures of PNA or Natura 2000 areas, as well as wetlands that lack such protection. In this scenario, only Galicia would take a step towards the protection of wetlands through a new conservation and management regime for wetlands that was established by Law 9/2001, introducing the category of "Protected Wetlands" within the different types of PNA. This new category was reserved for those wetlands that fulfilled a function of international, national, or regional importance in the conservation of natural resources, and that were declared as such.

Derived from these previous developments in terms of the regional legal framework and protection of wetlands, and in application of Law 4/1989, it was considered necessary to carry out an inventory of them at Galician level, which would allow identifying their territorial representativeness, their conservation status, as well as the role they played in conservation of biodiversity and the functionality of ecosystems. In this way, the scope of action of the first Galician Wetlands Inventory (GWI) was established [29], which was carried out under a collaboration agreement signed between the Galician government and the University of Santiago de Compostela over the years 2001 and 2003. The GWI scientific-technical aspects were later included in Decree 127/2008, whereby developed the legal regime of protected wetlands and created the Galician Wetlands Inventory (GWI).

4. The Galician Wetlands Inventory (GWI)

Galician Wetlands Inventory (GWI) included those natural, semi-natural or artificial systems that could be attributed to one of the types established in the classification of Ramsar wetlands and whose environmental interest could be corroborated with any of the internationally approved systems (Ramsar Convention, Birds Directive, Habitat Directive, IUCN) for the characterization of biodiversity at the level of its biotic components and the eco-functions they perform in the system.

The typology of the GWI integrated the criteria of the Ramsar Wetlands Classification established in Recommendation IV.7 approved by COP4 Montreux-1990 and modified by Resolution VI.5 approved by COP6 Brisbane-1996 [1], and the criteria of the Eunis-Habitat classification [38, 39], as well as the categories of the Spanish Strategic Plan for the Conservation and Rational Use of Wetlands [4]. For some categories (lagoon environments), standardized delimitations for the characterization of the different units (lakes, lagoons, ponds, puddles) were fixed. The typology of wetlands thus showed an easy correspondence with the one used by the European Commission for the designation of the types of habitats of community interest present in Annex I of DC 92/43/EEC.

As a result of the inventory and identification work, the GWI included a total of 1,131 wetlands distributed mostly between the coastal sectors, the interior sedimentary depressions and the sub-littoral and central mountain areas. The rest of the Galician territory and specifically the eastern and southern mountainous areas present a smaller proportion of humid ecosystems, which are usually confined to areas with morphological characteristics favourable to the maintenance of seasonal water contributions (**Figure 4**). In total they cover more than 70,600 ha (**Table 1**), which represents 2.4% of the Galician continental territory [29].

Comparing the data obtained in the GWI [29], with the information available in the rest of Spanish regions for the same inventory period (2001–2003), Galicia would be the region with the largest number (**Figure 3**), and probably the largest area of wetlands, having more than 26% of the Spanish wetlands. According to the major types of wetlands established in the Inventory of Wetlands of Galicia (**Table 1**), the majority of the inventoried means correspond to the group of "continental wetlands" with a total of 659 wetlands (54.4% of the total) followed by the "artificial wetlands" with 312 wetlands (27.6%) and "marine-coastal" with 153 wetlands (13.3%), while the number of "subterranean wetlands" would be represented by 7 wetlands.

The littoral Galicia includes the largest average size wetlands, as they usually appear forming important coastal and marine complexes, including shallow marine waters, humid dune slacks, estuarine systems, coastal lagoons, humid meadows, and alluvial forests [15–17]. In the interior of Galicia, the sedimentary basins and the bottoms of the great Atlantic valleys host an important representation of peaty, hygrophilic, fluvial and lacustrine wetlands [6, 30]. Among the large sedimentary basins, upper Miño basin stands out for the biodiversity, natural heritage, and naturalness of its ecosystems, as the floodplain configured around the main channels of the upper Miño river and tributaries includes one of the best representations of the

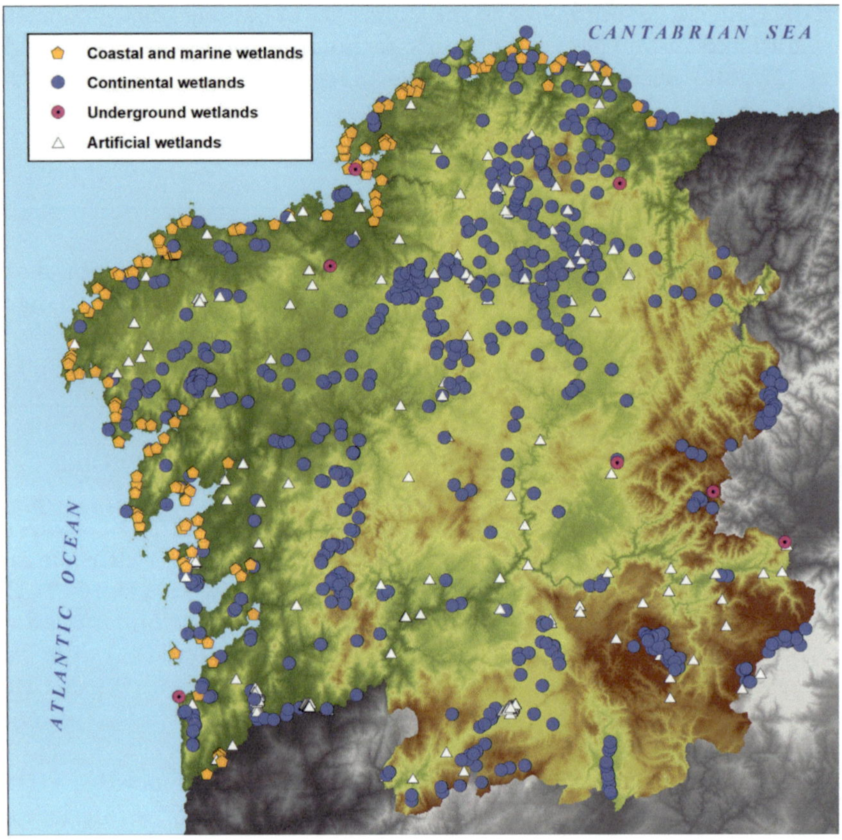

Figure 4.
Geographical distribution and typology of wetlands included in the GWI [29]. The symbols represent the centroid of the wetlands.

	Number		Area	
Mayor types of wetlands	**n**	**%**	**ha**	**%**
Coastal and marine wetlands	153	13.3%	20,268,2	28.7%
Continental wetlands	659	54.4%	34,442.7	48.8%
Underground wetlands	7	0.6%	0.6	<0.1%
Artificial wetlands	312	27.6%	15,965.8	22.7%
Total	1,131		70,677.3	

Table 1.
Major types of wetlands according to the classification established in the GWI [29].

Atlantic biogeographical region of alluvial forests, mixed with ponds, grasslands and wet scrub [34, 40].

The Galician mountains also include a significant number of wetlands. Their largest area is concentrated in bogs and wet heaths, standing out Serra do Xistral, where the most important blanket bog complex of Southern Europe is located [33]. On the contrary, the lakes located in the mountainous areas are very small, although numerous, and mostly originated in basins of glacial origin.

5. Known vs. protected wetlands

As the Ramsar Convention establishes, wetland inventorying is the basic tool for planning and carrying out wetland management, protection and conservation activities. In Spain, the SWI would hypothetically take the data collected in regional lists or inventories, which should be carried out and updated by regional governments, with the participation of different specialists and research centres. The inclusion of a wetland in the SWI (Royal Decree 435/2004) does not give it a protection status, as the inventory contributes to knowing the evolution of wetlands and, where appropriate, indicating the protection measures that must be included in the basin hydrological plans. It is only possible to entering the inventory those natural, semi-natural or anthropic environments that meet morphological and/or ecological characteristics defined in accordance with those set by Ramsar, or those that have expressly attributed the status of wetland by virtue of a specific protection regulation. These same criteria are followed by the Spanish regions in their own legal framework. Thus, in Galicia (Decree 127/2008), the inclusion of a wetland in the inventory must meet a series of scientific-technical criteria, which apply both to those designed under a specific protection category and to those not. In the Galician regulations, the inclusion of a wetland in the inventory is carried out for statistical and research purposes and does not automatically imply the application of a protection status.

Regarding the wetland protection in Spain, this is established through its recognition under different categories linked to land or water planning and management, or through its designation under one of the different categories of PNA, which may be common to the used ones for other ecosystem types (National Parks, Natural Monuments, Protected Landscapes), or more rarely one of those specific to wetlands, as occurs in Galician regulations with the creation of the figure of "Protected wetland", established by Law 9/2001.

It is assumed that the number of known wetlands that appear compiled in the scientific-technical inventory documents that have been carried out in the different Spanish regions should be equal to or slightly less than the registered wetlands in the regional official catalogues as well as than the registered wetlands in the SWI. Currently, the SWI includes 715 wetlands, a value that represents 14.3% of Spanish known wetlands. This meagre representation is even more worrying when their territorial distribution is analysed, since 583 (11.66%) correspond to wetlands distributed in the Spanish southern regions which receive less rainfall, like Madrid (23 wetlands), Castilla - La Mancha (312 wetlands), Andalusia (147 wetlands), Murcia (53 wetlands), or Comunitat Valenciana (48 wetlands). While in the Spanish northern regions, that receive a higher rainfall which usually offers better conditions for the configuration and persistence of wetlands, only 132 wetlands (2.64%) have been recorded in the SWI with a very unequal distribution across the northern regions, like Galicia (0 wetlands), Asturias (53 wetlands), Cantabria (0 wetlands), Euskadi (30 wetlands), Navarra (0 wetlands), La Rioja (49 wetlands), or Castilla y León (0 wetlands). Consequently, the SWI has excluded 1,895 known wetlands in Northern Spain, which means 93% of them, and 38% of all the Spanish known wetlands [29].

Regarding the protection status, in recent years there has been a significant adaptation and improvement of both Spanish and regional environmental regulations, improving their convergence with the criteria established in other different countries. The application of regulations and monitoring mechanisms for their compliance continue to be a pending issue that has a significant impact on the conservation status of wetlands and on the authorization and sanction regime for different activities and uses. In the application of some sectoral regulations, such as those referring to water, the vision of wetlands continues to focus on reservoirs and lotic and lentic ecosystems, while hygrophilous or turfophilic wetlands are marginalized [37]. This unequal treatment is clearly seen in the basin plans, where the inventory, characterization, and proposal for the wetlands and specifically the hygrophilous and turfophilic ones is very deficient.

The environmental regulations in Spain establish three large groups of protected areas since Law 42/2007 [41, 42]: PNA (Reserves, Parks, Natural Monuments, Protected Landscapes, Marine Protected Areas), Natura 2000 protected areas (SCI, SAC, SPA), and APII (Ramsar Wetlands, Global Geoparks, World Heritage, Biosphere Reserves, etc.). Natura 2000 are the protected areas that include more sites and occupy the greater surface of wetlands, usually overlapping SAC and SPA and thus strengthening the level of protection on the same wetland, as well as establishing important synergies with other categories of protected areas. As an example, Natura 2000 areas support the core zones of the 20 Biosphere Reserves that have been designed in Northern Iberian Peninsula, which sum to a maritime-terrestrial area of 1.6 million hectares, including a notable representation of marine, coastal, interior and mountain wetlands [43]. However, during the creation process of Natura 2000,

Group A. Sites containing representative, rare or unique wetland types	
Criterion 1	It contains a representative, rare, or unique example of a natural or near-natural wetland type found within the appropriate biogeographic region.
Group B. Sites of international importance for conserving biological diversity	
Criteria based on species and ecological communities	
Criterion 2	It supports vulnerable, endangered, or critically endangered species or threatened ecological communities
Criterion 3	It supports populations of plant and/or animal species important for maintaining the biological diversity of a particular biogeographic region.
Criterion 4	It supports plant and/or animal species at a critical stage in their life cycles, or provides refuge during adverse conditions.
Specific criteria based on waterbirds	
Criterion 5	It regularly supports 20,000 or more waterbirds.
Criterion 6	It regularly supports 1% of the individuals in a population of one species or subspecies of waterbird.
Specific criteria based on fish	
Criterion 7	It supports a significant proportion of indigenous fish subspecies, species or families, life-history stages, species interactions and/or populations that are representative of wetland benefits and/or values and thereby contributes to global biological diversity.
Criterion 8	It is an important source of food for fishes, spawning ground, nursery and/or migration path on which fish stocks, either within the wetland or elsewhere, depend.
Specific criteria based on other taxa	
Criterion 9	It regularly supports 1% of the individuals in a population of one species or subspecies of wetland-dependent non-avian animal species.

Table 2.
The Ramsar Sites criteria for identifying Wetlands of International Importance [44].

the European Commission considered in 2004 that the network was incomplete in the Spanish Atlantic Region, as several wetland habitat types of community interest were covered insufficiently, such as wet heaths (Nat-2000 4040*) raised bogs (Nat-2000 7110*), and transition mires (Nat-2000 7140), so it would be necessary to revise the network, proposing new sites or enlarging some of the proposed at that time.

The rest of PNA like the Parks (National Parks, Natural Parks) and other categories (Reserves, Natural Monuments, etc.) occupy a lesser extent than Natura 2000. Regarding APII, their protection regime is determined by the corresponding international conventions and agreements, without prejudice to the validity of specific protection, planning and management regimes whose territorial scope coincides totally or partially with these protected areas, if they comply with the provisions of such international instruments.

As for the APII category directly related to wetland protection, the Ramsar sites, there are currently in Spain 76 wetlands (more than 316,000 ha) included in the Ramsar list, that accomplish different Ramsar criteria (**Table 2**): 31.70% wetlands meet criterion 2 (they support vulnerable, endangered or critically endangered species, or threatened ecological communities), while 26.8% meet criterion 3 (they support populations of plant and/or animal species important to maintain the biological diversity of a biogeographical region), 14.6% meet criterion 1 (they contain a representative, rare or unique example of a natural or near-natural wetland type within a biogeographical region), and 11.6% meet criterion 6 (they regularly support 1% of the individuals of a population of a species or subspecies of waterfowl). Compliance with the rest of the criteria is below 10% (**Figure 5**).

Among the Spanish Ramsar sites, 19 of them are spread across the Spanish northern regions (**Table 3**): Galicia (5 wetlands +1 shared one), Asturias (1 wetland +1

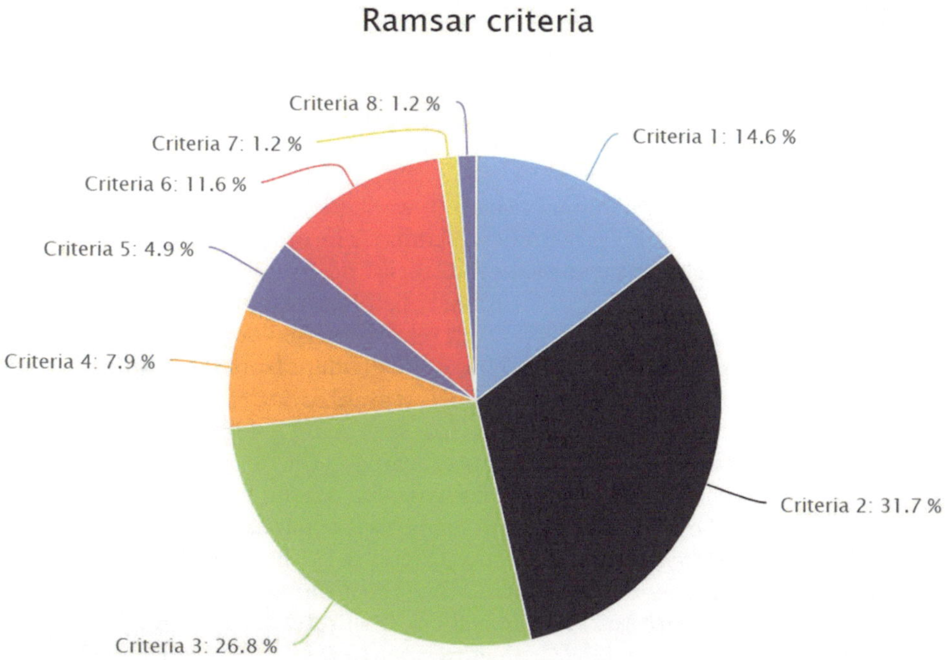

Figure 5.
Criteria met by Spanish wetlands included in the Ramsar List [44].

Ramsar site	Date	Area (ha)	Region	Type
Ría de Ortigueira y Ladrido	12/05/1989	2,600.58	Galicia	Littoral
Complexo intermareal Umia-Grove	12/05/1989	2,920.00	Galicia	Littoral
Lagunas de Villafafila	12/06/1989	2,714.00	Castilla y León	Lacustrine
Ría de Mundaka-Guernica	26/03/1993	945.00	Euskadi	Littoral
Lagoa e areal de Valdoviño	26/03/1993	485.23	Galicia	Littoral
Complexo de Corrubedo	26/03/1993	983.80	Galicia	Littoral
Ría de Ribadeo	04/10/1994	614.00	Galicia-Asturias	Littoral
Marismas de Santoña, Victoria y Yoyel	04/10/1994	6,907.00	Cantabria	Littoral
Lagunas de Laguardia	12/09/1996	45.00	Euskadi	Lacustrine
Embalse de las Cañas	18/11/1996	101.00	Navarra	Reservoir
Laguna de Pitillas	18/11/1996	216.00	Navarra	Lacustrine
Colas del Embalse de Ullibarri-Gamboa	24/10/2002	397.00	Euskadi	Reservoir
Lago de Caicedo-Yuso y salinas de Añana	24/10/2002	26.00	Euskadi	Lacustrine
Laguna de La Nava de Fuentes	24/10/2002	326.00	Castilla y León	Lacustrine
Salburua	24/10/2002	174.00	Euskadi	Lacustrine
Txingudi	24/10/2002	128.00	Euskadi	Littoral
Humedales de la Sierra del Urbion	19/02/2007	86.00	La Rioja	Mountain
Ría de Villaviciosa	07/01/2011	1,263.00	Asturias	Littoral
Galician Atlantic Islands	05/22/2021	8,542.62	Galicia	Littoral
Total		29,474.23		

Table 3.
Ramsar sites in the Spanish northern regions. Created by the authors from official Ramsar information.

shared one), Cantabria (1 wetland), Euskadi (6 wetlands), Navarra (2 wetlands), La Rioja (1 wetland) and Castilla León (2 wetlands). The only shared Ramsar wetland in this group, which is located between Galicia and Asturias regions, has been erroneously named as 'Ría del Eo' in the Ramsar list, but its right name should be 'Ría de Ribadeo', according to the most recent reports carried out by the Spanish Specialized Commission of Geographic Names (dependant from the Spanish Superior Geographical Council) and the Spanish Royal Geographic Society, following historical criteria. The rest of Ramsar wetlands in Northern Spain are framed in a single region each one. Ten of these Ramsar sites are located in the littoral area, with an unequal representation of littoral marine habitats and coastal habitats. The remaining 9 are in inland areas, 2 of them correspond to reservoirs, 6 to lacustrine environments and only 1 correspond to mountain wetlands (Sierra del Urbión Wetlands).

The Spanish environmental regulations allow the different Spanish regions to establish their own categories of PNA. In Galicia, the Protected Wetlands have been established as a proper regional PNA category since Law 9/2001 (setting an unparalleled example in Spain at that time), which was supposedly created to provide a specific regional protection regime for wetlands, promoting a better conservation of these ecosystems, taking into account their special fragility and value from an environmental point of view, and specifically dedicated for those declared as International Importance according to the Ramsar Convention. This regional category has been maintained by Law 5/2019, which in addition has incorporated the Ramsar Wetlands of International Importance into the Galician legal framework, within the group category of APII, inheriting the model previously established for the Spanish territory by Law 42/2007. However, nowadays this regional category of PNA has only been applied to the first 5 Ramsar sites designed in Galicia (**Table 3**), as the last one designed (Galician Atlantic Islands) in 2021 [45] does not have received yet the designation as a Protected Wetland, nor has the procedure for such designation been initiated, nor has a preventive protection regime for the site been established.

6. Conservation status of Spanish wetlands

The Iberian Peninsula is frequently considered as a territory of high biodiversity with a great diversity of habitats and species, considered endemic, rare or threatened with disappearance [35, 43, 46]. The Iberian high value of biodiversity contrasts with the unfavourable conservation status of many of the biodiversity components, which is caused by human action. In this scenario, wetlands are not an exception. Already in 1990, the Ramsar Convention established the "Montreux Registry", to include those Ramsar wetlands in which "modifications in ecological conditions have occurred, are occurring or may occur", so it should be necessary to carry out priority conservation actions. Among the worldwide 872 Ramsar sites, 62 are currently included in the Montreux Registry, and two of them are Spanish, having been included in this registry since 07/04/1990: Doñana (Andalucia) and Las Tablas de Daimiel (Castilla-La Mancha). Both wetlands are National Parks and have also been designed under other different categories of PNA, but they have registered numerous complaints filed in courts about their poor situation, as well as to the different international organizations that manage the different programs and categories of protected areas. Doñana was added to the Montreux Registry due to a situation generated by tourist pressure and intensive irrigation agriculture that are present in and around the wetland. Intensive irrigation agriculture is also the determining cause Las Tablas de Daimiel deterioration, as the groundwater intakes have caused a progressive drop in the water table, modifying the hydrological cycle that fed Guadiana River and Las Tablas de Daimiel wetland. Despite attempts to reverse the situation in both wetlands, their unfavourable situation determines that they continue to be included in the Montreux Registry.

The situation of other Spanish wetlands is equally worrying, despite not having been included in the Montreux Registry. Among them, Lagoa e Areal de Valdoviño (Galicia) Ramsar site accumulates many complaints [47] due to the change of its limits (**Figure 6**), to the alteration of the hydro-ecological cycle with rupture of the coastal barrier and alteration of the flooding regime, loss of area occupied by natural-semi-natural habitats, expansion of invasive alien species, irrational and unsustainable public use, etc. In other cases, human activities have led to the total destruction of the wetland, transforming the area into intensive agricultural crops or Eucalyptus plantations [37].

In recent works [37, 46, 48], the situation of Galician wetlands has been analysed for the last 20 years, comparing the situation at the starting of the works leading to the GWI (2001–2003), with the current situation. From them, it can be deduced that human action continues to have a very negative effect on the conservation of wetlands,

Figure 6.
Change of limits in Lagoa e Areal de Valdoviño Ramsar site, from its initial designation in 1992 and the subsequent modification of its boundaries in 2006, that has caused numerous complaints [47].

both in those cases in which they are protected under any category of protection, and more dramatically, in those wetlands that have not been designed as protected areas. Conservation status of Galician wetlands can be evaluated by applying the same methodology used for the Directive 92/43/EEC Article 17 reporting every six years for conservation status of habitats and species of community interest [49], focusing on the habitat types that are characteristic of the natural wetlands in Galicia [18, 20, 21, 29]. In each type of wetland were valued 4 factors: occupancy area, habitat structure, alterations on the hydrology, and conservation status of wetland characteristic species. For each one of them, different parameters were recorded (**Figure 7**).

The different impacts that negatively affect the conservation status of Galician wetlands were assessed following the systematization established in Spain [50], which follows the previous standardized list of the European Commission. Impacts are grouped in 4 classes (A.- Activities of the primary sector. B.- Construction activities. C.- Pollution. D.- Activities of public use), in which 4 types of actions are included. Among the 29 wetland types of the GWI that were evaluated (**Figure 7**), only 3 (10.35%) have a "Favourable" conservation status, which are underground habitats (3.1.1 and 3.2.1) and shallow marine waters (1.2.1). With a "Moderate" conservation status there is only one habitat (3.45%), that corresponds to estuaries (1.3.1). The wetlands in an "Inadequate" conservation status are 8 types (27.58%), including sandy coastal systems (1.2.3, 1.4.2), intertidal flats (1.2.4, 1.3.2), intertidal marshes (1.3.3),

Wetland typology			CS				A				B				C				D			
GWI	**R**	**Denomination**	1	2	3	4	1	2	3	4	1	2	3	4	1	2	3	4	1	2	3	4
1.	--	***Marine and coastal wetlands***																				
1.2.	--	***Shallow marine waters***																				
1.2.1	B	Permanent marine waters	-	-	-	+	+	-	-	-	-	-	-	-	+	●	-	-	-	-	-	-
1.2.3	E	Coastal sandy systems	+	+	-	+	+	-	-	-	+	-	-	+		●	-	-	●	-	-	●
1.2.4	G	Intertidal mud or sand	+	+	-	+	+	-	-	-	+	-	-	-	+	●	-	-	+	-	-	●
1.2.5	H	Intertidal marshes	+	+	-	●	+	+	+	-	+	-	-	-	+	●	+	+	+	+	+	●
1.3.	--	***Estuaries and fluviomarine systems***																				
1.3.1	F	Estuarine waters	+	-	-	+	+	-	-	-	+	-	-	-	+	●	+	-	+	-	-	●
1.3.2	G	Intertidal mud or sand	+	+	-	+	+	-	-	-	+	-	-	-	+	●	+	-	+	-	-	●
1.3.3	H	Intertidal marshes	+	+	-	+	+	-	-	-	+	-	-	-	+	●	+	+	+	+	+	●
1.4.	--	***Coastal lagoons***																				
1.4.2	E	Coastal sandy systems	+	+	-	+	+	-	-	-	-	-	-	+	-	●	+	-	●	-	-	●
1.4.3	J	Coastal brackish/saline lagoons	●	+	●	●	+	+	+	-	-	-	-	+	+	●	+	+	●	+	+	●
1.4.4	K	Coastal freshwater lagoons	+	+	●	●	-	+	+	-	-	-	-	+	+	●	+	+	●	+	+	●
2.	--	***Continental wetlands***																				
2.1.	--	***Freshwater ecosystems***																				
2.1.1	M	System of terrigenous islands	+	●	+	+	-	+	+	-	-	-	-	+	+	●	+	-	●	-	+	●
2.1.2	M	Cords of tree islands	●	●	●	-	-	+	+	-	+	-	-	-	+	●	-	-	●	-	-	-
2.1.3	M	Meanders temporarily unconnected	+	●	+	-	-	+	-	-	-	-	-	-		●	+	-	-	-	-	-
2.1.4	M	Waterfalls and wells.	+	+	+	-	-	+	-	-	-	-	-	-	+	●	-	-	●	-	-	-
2.2.	--	***Continental lacustrine ecosystems***																				
2.2.1	O	Permanent freshwater lakes	-	-	●	●	-	-	-	-	-	-	-	-	+	●	●	+	●	-	-	●
2.2.2	Tp	Lagoons (surface 1 - 8 ha)	+	●	●	●	-	-	-	-	-	-	-	-	+	●	●	+	●	-	+	●
2.2.3	Ts	Ponds (< 1ha)	+	●	●	●	-	●	+	-	+	-	-	-	+	●	●	+	●	-	+	●
2.2.4	Ts	Temporary lagoons or ponds	●	●	●	●	-	●	●	-	+	-	-	-	+	●	●	+	●	-	+	●
2.2.5	Ts	Ponds and seasonal puddles	●	●	●	+	-	●	●	-	+	-	-	-	+	●	●	+	●	-	+	●
2.3	--	***Non-forested peatlands***																				
2.3.1	U	Blanket bogs	●	●	+	+	-	●	●	●	●	-	●	●	-	-	●	●	-	-	+	-
2.3.2	U	Raised bogs	●	●	+	+	-	●	●	●	●	-	●	●	+	●	●	●	-	+	+	+
2.4	--	***Shrub-dominated wetlands***																				
2.4.1	W	Wet heaths	●	●	+	+	-	●	●	●	●	-	●	●	-	●	●	●	-	+	+	+
2.4.2	W	Wet shrubland	-	+	+	+	-	-	-	-	-	-	-	-	-	-	●	-	-	+	-	+
2.6.	--	***Hygrophilic ecosystems***																				
2.6.2.1	Xf	Freshwater tree-dominated wetlands	●	●	+	●	-	●	●	+	●	-	-	-	+	●	+	+	-	-	-	-
2.6.2.2	W	Shrub-dominated swamps/fens	●	●	+	+	-	+	+	-	●	●	●	-	-	●	+	+	-	+	+	+
2.6.2.3	Tp	Permanent freshwater swamps/fens	-	+	+	+	-	-	-	-	-	-	-	-	-	-	-	-	-	-	-	-
2.6.2.4	Ts	Seasonal freshwater swamps/fens	+	●	+	+	-	+	+	-	●	-	-	-	-	●	+	-	-	-	-	-
3.	--	***Underground water systems***																				
3.1.		***Marine underground wetlands***																				
3.1.1.	Zk (a)	Marine/coastal subterranean systems	-	-	-	+	+	-	-	-	-	-	-	-	+	●	-	-	-	-	-	-
3.2.		***Inland underground wetlands***																				
3.2.1.	Zk (b)	Inland subterranean systems	-	-	-	+	+	-	-	-	-	-	-	-	+	●	-	-	-	-	-	-

[CS] – Conservation status indicators. [1] Occupancy area. [2] Structure. [3] Hydrology. [4] Species. Non-significant or non-evaluable condition (-). Significant impact (✚). Very significant impact (●). Conservation status of wetlands: **Critical**: 4● │ 3● and 1✚. **Bad**: ≥ 2● │1 ● and ≥1✚ │1● and 3✚; **Inadequate:** 1 ● │ ≥3✚; **Moderate**: ≥2✚; **Favourable**: Rest of combinations.

Factors affecting the conservation status and biodiversity loss of wetlands: (A) Activities of the primary sector. [A1] Fishing-seafood activity. [A2] Agricultural-livestock activity. [A3] Forestry activity. [A4] Extractive activity. (B) Construction activities: [B1] Roads and linear infrastructures. [B2] Industrial, commercial, or service areas. [B3] Renewable energy installations (wind farms). [B4] Buildings and isolated constructions. (C) Pollution: [C1] Pollution of urban - industrial origin. [C2] Diffuse contamination. [C3] Use of lead ammunition in wetlands. [C4] Noise pollution. (D) Public use activities: [D1] Pressure on ecologically sensitive areas. [D2] Collection of natural components, accidental spread of IAS. [D3] Circulation - parking of vehicles. [D4] Uncontrolled presence of pets (without leash or muzzle).

[GWI]: Galician Wetland Inventory typologies. [R]: Ramsar Convention types of wetlands.

Figure 7.
Conservation status of Galician wetlands, comparing between 2001 and 2022. Created by the authors from recent works assessing the situation of Galician wetlands [37, 46, 48].

waterfalls (2.1.4), humid shrubland (2.4.2), and permanent freshwater swamps/fens (2.6.2.3). In a "Bad" conservation status there are 11 types of humidais (37.93%): Intertidal marshes (1.2.5), Coastal freshwater lagoons, (1.4.4), riverine tree and terrigenous islands (2.1.1, 2.1.2), temporarily disconnected meanders (2.1.3), permanent freshwater lakes (2.2.1), shrub-dominated and seasonal freshwater swamps/fens (2.6.2.2, 2.6.2.4), blanket and raised bogs (2.3.1, 2.3.2), and wet heaths (2.4.1). And finally, in a "Critical" conservation status, 6 types of wetlands are included (20.69%): coastal brackish/saline lagoons (1.4.3), permanent, temporary, and seasonal continental lagoons and ponds (2.2.2, 2.2.3, 2.2.4, 2.2.5), and freshwater tree-dominated wetlands (2.6.2.1).

Anthropogenic actions that have a very negative effect on the conservation status of wetlands are mostly linked to actions derived from construction activities, as well as those linked to the primary sector. Their execution usually leads to the partial loss of habitats that are characteristic of wetlands, but even sometimes they also generate the complete destruction of the wetland [37]. Thus, permanent ponds within Parga-Ladra-Támoga SAC (ES1120003) have been lost because of their transformation into intensive farmland, a situation that similarly happened in the coastal lagoon of Praia da Ermida (Costa da Morte SAC, ES1110005), where the wetland disappeared during the execution of supposedly environmental restoration works. In other cases, it is possible to identify large areas of wetlands that have been transformed by the afforestation with exotic species (*Eucalyptus, Pinus*) or by the creation of artificial pasturelands, as it can be documented in different areas of Serra do Xistral SAC (ES1120015), causing a high impact on peaty habitats like blanket bogs, raised bogs and wet heaths [33].

The installation of wind farms also had a very negative effect on mountain wetlands, causing significant losses [36] in wet heaths, peaty ecosystems (blanket bogs, raised bogs), as well as on lacustrine environments. Serra do Xistral SAC (ES1120015) also stands out regarding the impact of wind farms on natural ecosystems [51], but this can also be detected in other areas of Galician Natura 2000 network. The negative effects of the installation of wind farms on the mountain wetlands of Galicia were pointed out when this activity erupted with force during the late 1990s in the region [33, 52]. Years later, there is no doubt about the high impact that windfarm developments generate on mountain areas, especially considering that they are established on the areas of most fragility and highest ecological value of the territory, as different authors have highlighted both for Galicia and for other territories in Europe [53–60]. Other reasons of wetland loss join to the previous ones, caused by the exploitation of subterranean resources, highlighting the extraction of fine and coarse aggregates on different Galician wetlands, as well as the peat extraction in Montes do Buio (again in Serra do Xistral SAC, ES1120015), to be used as a substrate for gardening.

The physic-chemical and microbiological analysis carried out in numerous wetlands in Galicia highlight the treatment problems of both urban-industrial waste water and that derived from livestock farms [37, 46]. The recurring presence of *Escherichia coli* and other pathogens of stool origin in coastal and inland wetlands can hardly be assimilated to a favourable ecological status. Regarding other chemical pollutants (antibiotics, phytocides), there is no detailed information that allows us to assess their impact on Galician wetlands, but in any case, their impact on the biodiversity components, especially amphibian populations, should not be minimized. The presence of lead ammunition in the Galician wetlands is more frequent than expected, despite its possession and use in wetlands it is considered illegal by Spanish law, but it has been diluted in Galician regulations, avoiding spreading among hunters the ban that affects this ammunition made with a toxic element [37].

Finally, it is necessary to comment on the impacts derived from actions of public use and leisure activities developed with disrespectful behaviour towards the values of natural heritage and biodiversity, derived from the banalization of natural areas as a mere tourist attraction or as an amusement park. This conceptualization is often favoured by tourism campaigns promoted from public bodies (mostly regional and local governments). Irrational public use is the cause of the deterioration of habitats and protected species populations, especially in the coastal wetlands, among which the most frequent actions are the usual presence of pets without leash and muzzle that freely roam into wildlife breeding or refuge areas of wetlands, the circulation and parking of vehicles on natural and semi-natural habitats, the collection of key components of the natural heritage and biodiversity, or the accidental spread of invasive alien species [37].

7. Future challenges of protected wetlands

Safeguarding of wetlands in Northern Iberian Peninsula requires a change of attitude towards natural heritage and biodiversity, abandoning the usual reactive policies and actions inherited from the old pre-democratic regimes, and start to develop proactive actions, facing the future challenges that arise over their protection. The current worldwide and European framework on biodiversity constitutes an effective action framework to reformulate and ensure the conservation of natural heritage and biodiversity of wetlands, as well as to improve the benefits they provide for people. According to Aichi Biodiversity Targets [61], at least 17% of terrestrial and inland water, and 10% of marine areas, should be protected. The EU Biodiversity Strategy for 2030 [62] increases this percentage to 30%, both in the marine and terrestrial areas. Northern Iberian Peninsula is very far from the reference values contemplated at international and European level, and it is aggravated considering several regions such as Galicia, where the terrestrial PNA do not even reach 13% [41, 42].

So, it is appropriate to raise the urgent need to increase the representation of wetlands in the different categories of protected areas (PNA, Natura 2000, APII), proposing new sites or enlarging some of the existing ones that should ensure adequate representation and protection of those wetlands that have habitats and/or endemic, rare or threatened species, and that play a significant role as a medium or long-term carbon reservoir. These usually correspond in Northern Iberian Peninsula to the different types of peat bogs, wet heaths, lagoons, and marshes. Similarly, it is equally necessary to add certain wetlands to the Ramsar List, trying to cover the current existing huge gap in relation to this category in the Cantabrian-Atlantic mountainous areas, as well as in the wetlands located in inland areas, both in Galicia (Terra Chá) and in the Iberian Plateau. Anyway, all the proposals for enlargement or new designation of wetlands protection in Galicia should obviously be complemented with measures for their mandatory monitoring, restoration and surveillance, as well as with the effective application of the sanctioning and criminal regime for every activity, plan or project that could carry out actions on wetlands that negatively affect their ecological integrity, or the conservation status of habitats and species held in them, and therefore contradict the laws and other general provisions on nature or environment. This approach assumes:

1. Eliminate or reduce close to zero the negative effects that are currently occurring in protected areas on the conservation status of habitats and species that are characteristic of wetlands. This would force to increase surveillance with rangers and complement it with remote sensing monitoring. Changes in use, waste disposal, mechanical shredding vegetation, the use of lead ammunition, or even the use of fire and biocides, which occurs in many wetlands, are incompatible with the objective of ensuring a favourable ecological status of these ecosystems.

2. Among the Spanish wetlands, there is a group of them that have lost their naturalness due to human action and that should be the object of restoration and conservation measures, to reverse this situation fully or partially. This list usually includes very large wetlands that were altered during Franco's dictatorship regime (Antela or Cospeito in Galicia, Mar de Campos in Castilla y León). But those wetlands whose significant alteration has occurred in recent decades should be added to this group, as they will end up disappearing if effective measures are not taken to ensure their conservation (being possible to cite in Galicia several examples: Brañas do Deo, Brañas de Brins, Brañas de Alcaián, Brañas de Xestoso, Lagoa de Valdoviño, etc.).

3. Surveillance, control, and eradication of invasive alien species must be done, acting on the mechanisms that facilitate their expansion, and therefore significantly reducing the area occupied by them, especially on natural and semi-natural habitats, whose ecological restoration should be promoted. The future scenarios marked by climate change are very negative for the conservation of wetlands, especially in the Atlantic coastal and insular areas of the Northern Iberian Peninsula, where the predicted scenarios will be favourable for the invasive alien species already installed in the territory, as well as for others that may settle soon.

4. The rational and sustainable management of wetlands must be based on the best scientific-technical information available and on previous experiences. It is anomalous to propose management actions aimed at favouring livestock use in mountain wetlands with actions that involve a significant alteration of the structure and composition of the natural and semi-natural habitats held in these wetlands, as is occurring in certain mountainous enclaves, such as Serra do Xistral SAC (ES1120015), where the increasing livestock pressure and the adoption of inappropriate use measures are negatively affecting the conservation status of peat bogs and wet heaths.

Acknowledgements

This study was supported by the LIFE Programme (which is managed by the European Climate, Infrastructure and Environment Executive Agency, CINEA) of the European Commission, with a EU funding contribution of 75% (project reference: LIFE20 NAT/ES/001007). This work only reflects the authors' view, exempting CINEA and European Commission for any use that may be made of the information it contains. The authors thank Galician Atlantic Islands National Park for providing the data regarding the declaration of the Ramsar site.

Author details

Javier Ferreiro da Costa* and Pablo Ramil-Rego
GI-1934-TB, Institute of Agricultural Biodiversity and Rural Development (IBADER), University of Santiago de Compostela, Lugo, Spain

*Address all correspondence to: javier.ferreiro.dacosta@gmail.com

References

[1] Ramsar Convention Secretariat 1971. The Ramsar Convention Manual: A Guide to the Convention on Wetlands (Ramsar, Iran, 1971). 4th ed. Gland: Ramsar Convention Secretariat; 2006. p. 114

[2] European Environment Agency (EEA). Progress towards Halting the Loss of Biodiversity by 2010. Luxembourg: Office for Official Publications of the European Communities; 2006

[3] Ministerio de Medio Ambiente. Estrategia Española para la Conservación y el uso Sostenible de la Diversidad Biológica. Madrid: Secretaría General de Medio Ambiente. Dirección General de Conservación de la Naturaleza. Ministerio de Medio Ambiente (MMA); 1998

[4] Ministerio de Medio Ambiente. Plan Estratégico Español para la Conservación y el Uso Racional de los Humedales, en el marco de los ecosistemas acuáticos de que dependen. Madrid: Ministerio de Medio Ambiente (MMA); 1999. p. 95

[5] Ramsar Convention Secretariat. Ramsar Convention's first Strategic Plan 1997-2002. In: Resolution VI.14. 6th Meeting of the Conference of the Contracting Parties (COP6). Brisbane: Ramsar Convention Secretariat; 1996

[6] Ramil-Rego P, Domínguez Conde J. A Lagoa de Cospeito. Historia e vida dun humidal chairego. Santiago de Compostela: Xunta de Galicia; 2006

[7] Rodríguez Guitián MA, Ramil-Rego P. Fitogeografía de Galicia (NW Ibérico): análisis histórico y nueva propuesta corológica. Recursos Rurais. 2008;**4**:19-50

[8] Mata Olmo R, Sanz HC. Atlas de los paisajes de España. Madrid: Ministerio de Medio Ambiente; 2004

[9] Pérez Alberti A, Borobio Sanchis M, Castillo-Rodríguez F, Payan-Pérz M. Metodología y clasificación de tipos de paisaje en Galicia. Revista de Geografia e Ordenamento do Território (GOT). 2014;**6**:259-282

[10] Ramil-Rego P, Rodríguez Guitián MA, Rubinos Roman MA, et al. La expresión territorial de la biodiversidad. Paisajes y Hábitats. Recursos Rurais, Serie Cursos. 2005;**2**:109-128

[11] Cobo Gradín F. Proyecto Galicia. Natureza. Zooloxía Tomos 38-40. A Coruña: Hércules Edicións; 2002

[12] Diaz-Fierros F. Proyecto Galicia. Natureza. Xeoloxía Tomos 36-37. A Coruña: Hércules Edicións; 2002

[13] Rigueiro Rodríguez A. Proyecto Galicia. Natureza: Botánica Tomos 41-43. A Coruña: Hércules Edicións; 2002

[14] Vieitez Cortizo E, Rey Salgado JM. A natureza ameazada 2004. Santiago de Compostela: Consello da Cultura Galega; 2005

[15] Ramil-Rego P, Rodríguez Guitián MA et al. Reseña del Patrimonio Natural y la Biodiversidad de Galicia: año 2005. Monografías do IBADER - Serie Biodiversidade. Lugo: IBADER, Universidade de Santiago de Compostela; 2005. p. 1070

[16] Ramil-Rego P, Rodríguez Guitián MA et al. Reseña do Patrimonio Natural e a Biodiversidade de Galicia: año 2008. Monografías do IBADER - Serie Biodiversidade. Lugo: IBADER, Universidade de Santiago de Compostela; 2008. p. 740

[17] Ramil-Rego P, Rodríguez Guitián MA et al. Reseña do Patrimonio

Natural e a Biodiversidade de Galicia: año 2012. Monografías do IBADER - Serie Biodiversidade. Lugo: IBADER, Universidade de Santiago de Compostela; 2012. p. 661

[18] Ramil-Rego P, Crecente MR. Plan Director da Rede Natura 2000 de Galicia. Documento Técnico. Santiago de Compostela: Dirección Xeral de Conservación da Natureza, Conselleria do Medio Rural (Xunta de Galicia) & Instituto de Biodiversidade Agraria e Desenvolvemento Rural, IBADER (USC); vol. 8. 2012

[19] Bañón Díaz R. Inventario de la biodiversidad marina de Galicia. Proyecto Lemgal. Santiago de Compostela; Xunta de Galicia. Consellería do Mar; 2017

[20] Ramil-Rego P, Rodríguez Guitián MA, Hinojo Sánchez BA, Rodríguez González PM, et al. Os Hábitats de Interese Comunitario en Galicia. Descrición e Valoración Territorial. Monografías do Ibader. Lugo: Universidade de Santiago de Compostela; 2008

[21] Ramil-Rego P, Rodríguez Guitián MA, Ferreiro da Costa J, Rubinos Román M, Gómez-Orellana L, de Nóvoa FB, et al. Os Hábitats de Interese Comunitario en Galicia. Fichas descritivas. Monografías do Ibader. Lugo: Universidade de Santiago de Compostela; 2008

[22] Ramsar Convention Secretariat. Wetland Inventory: A Ramsar Framework for Wetland Inventory and Ecological Character Description. 4th ed. Gland: Ramsar Convention Secretariat; 2010

[23] Pardo L. Catálogo de los lagos de España. Instituto Forestal de Investigaciones y Experiencias. Madrid: Ministerio de Agricultura; 1948

[24] Cirujano BS. Criterios botánicos para la valoración de las lagunas y humedales españoles (Península Ibérica y las Islas Baleaeres). Madrid: Ministerio de Agricultura, Pesca y Alimentación; 1992

[25] Troya Panduro A, Bermues SM. Humedales españoles en la lista del Convenio Ramsar. Madrid: Ministerio de Agricultura, Pesca y Alimentación. ICONA; 1990

[26] Bermues SM. Humedales Españoles inscritos en la Lista del Convenio de Ramsar. Madrid: Ministerio de Medio Ambiente; 1998

[27] Cirujano Bracamonte S, Velayos Rodríguez M, Castilla Lattke F, Gil PM. Flora y vegetación de las lagunas y humedales de la provincia de Cuenca. Madrid: Consejo Superior de Investigaciones Científicas; 1995

[28] Casado de Otaola S, Montes del Olmo C. Guía de los lagos y humedales de España. Madrid: Editorial Reyero, J.M; 1995

[29] Ramil-Rego P, Izco J. Inventario dos Humidais de Galicia. Memoria Técnica. Santiago de Compostela: Dirección Xeral de Conservación da Natureza, Conselleria de Medio Ambiente (Xunta de Galicia) & Laboratorio de Botánica e Bioxeografía (Universidade de Santiago); 2003

[30] Ramil-Rego P, Rodriguez Guitian M, Gomez-Orellana L, Munoz Sobrino C, et al. Biogeografia Pleistocena-Holocena de la Peninsula Iberica. Santiago de Compostela: Xunta de Galicia; 1996. pp. 227-246

[31] Ramil-Rego P, Rodríguez Guitián MA, Rodríguez-Oubiña J. Valoración de los humedales continentales del NW Ibérico: caracterización hidrológica, geomorfológica y vegetacional de las turberas de las Sierras Septentrionales de Galicia. In: Pérez Alberti A, Martínez Cortizas A, coordinators. Avances en la

reconstrucción paleoambiental de las áreas de montaña lucenses. Monografías G.E.P. nº1. Lugo: Diputación Provincial de Lugo; 1996. pp. 166-187

[32] Ramil-Rego P, Rodríguez Guitián MA, Muñoz Sobrino C. Distribución, génesis y caracterización botánica de las turberas ombrotróficas de Galicia. XII Bienal de la Real Sociedad Española de Historia Natural. 1996; Tomo extraordinario: 253-256

[33] Izco J, Ramil-Rego P. Análisis y valoración de la Sierra de O Xistral: un modelo de aplicación de la Directiva Hábitat en Galicia. Santiago de Compostela: Consellería de Medio Ambiente, Xunta de Galicia;2001; 162

[34] Ramil-Rego P, Ferreiro da Costa J. Biodiversidad del corredor fluvial del río Miño: tramo Ponte Ombreiro- Caneiro do Anguieiro (Lugo). En: Crecente Maseda M, González Soutelo S, editors. Dos mil años del balneario de Lugo. Un modelo de activación del patrimonio termal. Lugo: Crecente Asociados; 2016. pp. 65-97

[35] Ferreiro da Costa J, Ramil-Rego P. Biological Conservation and Nature Protection Strategies in Spanish Atlantic Region. In: Şen B, editor. Selected Studies in Biodiversity. Rijeka: InTech; 2018. p. 44

[36] Ferreiro da Costa J, Ramil-Rego P, Hinojo Sánchez B, Cillero Castro C, Rubinos Román M, Gómez-Orellana L, Diaz Varela RA. Diagnóstico y Caracterización de los Brezales Húmedos (Nat-2000 4020*) de las Sierras Septentrionales de Galicia a partir de Criterios Científicos: Importancia para su Conservación. Recursos Rurais. 2013;9:65-77

[37] Ramil-Rego P, Ferreiro da Costa J, Rodríguez Guitián MA, López Castro H, Gómez-Orellana L. Perdas e alteración da biodiversidade nos humidais de Galicia. In: Ramil-Rego P, Gómez-Orellana L, Ferreiro da Costa J, editors. Conservación e xestión de humidais en Galicia. Lugo: Horreum-Ibader; 2017. p. 127-167

[38] Davies CE, Moss D. EUNIS Habitat Classification. 2001 Work Programme. Final Report. Huntingdon: European Environment Agency; 2002; 108

[39] Davies CE, Moss D. EUNIS Habitat Classification. Marine Habitat Types: Proposals for Revised Criteria, July 2004. Report to the European Topic Centre on Nature Protection and Biodiversity. Huntingdon: European Environment Agency; 2004

[40] Ramil Rego P, Crecente Maseda R, Rodríguez Guitián MA, Rubinos Román M, de Nóvoa Fernández B, Hinojo Sánchez B, Ferreiro da Costa J, Cillero Castro C, Díaz Varela R, Martínez Sánchez S, Gómez-Orellana Rodríguez L, García Abad F. Alto Miño, Terra Chá. Santiago de Compostela: Fundación Comarcal Terra Chá. Agader. Xunta de Galicia; 2000. p. 157

[41] Ramil-Rego P, Ferreiro da Costa J, Gómez-Orellana L, López Castro H, Oreiro Rey C, Rodríguez Guitián MA. Áreas Naturales Protegidas, de las propuestas pioneras a los nuevos paradigmas en pro de la salvaguarda de la Naturaleza. Monografías do IBADER. Serie Biodiversidade. Lugo: Ibader. Universidade de Santiago de Compostela; 2021. p. 918

[42] Ferreiro da Costa. Las áreas protegidas de Galicia: análisis histórico y perspectivas de gestion [thesis]. Lugo: IBADER. Universidade de Santiago de Compostela; 2022

[43] Ferreiro da Costa J, Ramil Rego P. Biosphere Reserves of the Spanish Atlantic region: protected areas for the conservation of biodiversity and

sustainable development. In: Miranda Barrós D, editor. LAND MATTERS. Taking Stock and Looking Ahead. Santiago de Compostela: Universidade de Santiago de Compostela; 2021. pp. 173-210

[44] Ramsar Secretariat. Ramsar Sites Information Service [Internet]. 2022. Available from: https://rsis.ramsar.org/ris-search [Accessed: October 1, 2022]

[45] Ferreiro da Costa J, Ramil-Rego P, Rodríguez Guitián MA, López Castro H, Oreiro Rey C, Gómez-Orellana L, et al. Galician Atlantic Islands National Park: Challenges for the Conservation and Management of a Maritime-Terrestrial Protected Area. In: Suratman MN, editor. Protected Area Management - Recent Advances. London: IntechOpen; 2022. p. 23

[46] Ramil-Rego P, Ferreiro da Costa J, Gómez-Orellana L, Rodríguez Guitián MA. Humidais de Galicia: Inventario e valoración ambiental no período 2001-2016. In: Ramil-Rego P, Gómez-Orellana L, Ferreiro da Costa J, editors. Conservación e xestión de humidais en Galicia. Lugo: Horreum-Ibader; 2017. pp. 1-29

[47] Muñiz A. A Lagoa da Frouxeira, en destrución. CERNA. 2016;**76**:14-17

[48] Ramil-Rego P, Rodríguez Guitián MA. Hábitats de turbera en la Red Natura 2000. Diagnosis y criterios para su conservación y gestión en la Región Biogeográfica Atlántica. LIFE Tremedal. LIFE11 NAT/ES/000707. Lugo: Horreum-Ibader; 2017. p. 427

[49] European Commission. The State of Nature in the EU. Conservation status and trends of species and habitats protected by the EU Nature Directives 2013-2018. Luxembourg: Office for Official Publications of the European Union; 2021

[50] Comisión Estatal del Patrimonio Natural y la Biodiversidad. Directrices para la vigilancia y evaluación del estado de conservación de las especies amenazadas y de protección especial. Comité de Flora y Fauna Silvestres de la Comisión Estatal para el Patrimonio Natural y la Biodiversidad (CEPNB). Aprobadas por la Comisión Estatal para el Patrimonio Natural y la Biodiversidad Madrid, 18/12/2012. Madrid: Ministerio de Agricultura, Alimentación y Medio Ambiente; 2012

[51] Gómez-Orellana L, Hinojo Sánchez B, Rubinos Román M, Ramil-Rego P, et al. El sistema de turberas de la sierra de O Xistral como reservorio de carbono, valoración, estado de conservación y amenazas. Bol. R. Soc. Esp. Hist. Nat. Sec Geol. 2014;**108**:5-17

[52] Ramil-Rego P, Ramil-Rego E. Valoración del patrimonio natural e histórico de las sierras septentrionales de Galicia. Vilalba: Grupo de Estudios Paleoambientales (G.E.P.) & Museo de Prehistoria y Arqueología de Villalba; 1995. p. 214

[53] Gómez-Orellana L, Rubinos Román M, Cillero Castro C, Hinojo Sánchez B, Ramil-Rego P. Ferreiro da Costa J. Los humedales de Galicia como sumidero de carbono: evaluación, distribución y estado de conservación. Bol. R. Soc. Esp. Hist. Nat. Sec. Geol. 2014;**108**:19-26

[54] Atienza JC, Martín Fierro I, Infante O, Valls J, Domínguez J. Directrices para la evaluación del impacto de los parques eólicos en aves y murciélagos (versión 3.0). Madrid: SEO/BirdLife; 2011. p. 115

[55] Bright JA, Langston RW, Bullman R, Evans RJ, Gardner S, Pearce-Higgins J, et al. Bird Sensitivity Map to Provide Locational Guidance for Onshore Wind Farms in Scotland. Sandy: The Royal Society for the Protection of Birds; 2006. p. 136

[56] Lindsay R, Bragg O. Wind Farms and Blanket Mires: The Bog Slide of 16th October 2003 at Derrybrien, Co. Galway, Ireland. Gort: School of Health and Biosciences, University of East London; 2004. p. 135

[57] Pearce-Higgins JW, Stephen L, Langston RHW, Baibridge IP, Bullman R. The distribution of breeding birds around upland wind farms. Journal of Applied Ecology. 2009;**46**:1323-1331

[58] Wawrzyczek J, Lindsay R, Metzger MJ, Quétier F. The ecosystem approach in ecological impact assessment: Lessons learned from windfarm developments on peatlands in Scotland. Environmental Impact Assessment Review. 2018;**72**:157-165

[59] Armstrong A, Burton RR, Lee SE, Mobbs S, Ostle N, Smith V, et al. Ground-level climate at a peatland wind farm in Scotland is affected by wind turbine operation. Environmental Research Letters. 2016;**11**(4):044024

[60] Chico G, Clutterbuck B, Clough J, Lindsay R, Midgley NG, Labadz JC. Geo-hydromorphological assessment of Europe's southernmost blanket bogs. Earth Surface Processes and Landforms. 2020;**45**:2747-2760

[61] CBD. The Strategic Plan for Biodiversity 2011-2020 and the Aichi Biodiversity Targets: Living in Harmony with Nature. UNEP/CBD/COP/DEC/X/2. Decision Adopted by the Conference of the Parties to the Convention on Biological Diversity at Its Tenth Meeting. Nagoya: Convention on Biological Diversity; 2010

[62] European Commission (EC). Communication from the Commission to the European Parliament, the Council, the European Economic and Social Committee and the Committee of the Regions. EU Biodiversity Strategy for 2030. Bringing nature back into our lives. COM (2020) 380 final. Brussels: European Commission; 2020

Chapter 3

Impacts of Stone Quarrying on Local Vegetation in Mount Korok Area, Juba, Central Equatoria State, South Sudan

Pasquale Tiberio Moilinga and Makuac Robert Athian

Abstract

This study was carried out in three quarry sites at Mount Korok, also known as Jebel Kujur area, which is located within Juba Town Council in Central Equatoria State, South Sudan. The main aim was to assess the impact of quarrying activities on the local ground cover vegetation, mainly grasses and low-lying non-woody herbaceous plants. The methods used included, besides direct observations, iron frame quadrats of $1 \times 1\ m^2$ in size, for random sampling of attributes and community characteristics of the plants in three different sites. The first site was an old, abandoned stone-quarrying site; the second was where quarrying work was actively going on at the time of the study; and the third was an area never before exposed to stone quarrying (hence, acting as a control). Data were analyzed using descriptive statistics such as frequency distribution, density measures, diversity indices, and correlations. The research was carried out during the wet season when most plants were green and at different stages of flowering and/or fruiting from July through September, 2020. The results revealed that over 44 species of ground cover plants were identified, some of which were more abundant and had the widest distribution and frequency in the three study sites, including *Cynodon lemfuensis, Cyperus rotundus, Bracharia ramose, Merremia pinata, Cyanodon dactylon*, and *Digitaria fernatad*, whereas others were limited to one site or the other. Results also indicated that though stone-quarrying activities have impacts on ground cover plants, however, they are not the only factor affecting ground cover plants. More than 80% of the impacts on ground cover plants are caused by factors other than stone quarrying but were not identified during this study. It was therefore recommended that future studies in the area on the same theme should isolate the effects of stone quarrying on ground cover plants from these other operating factors through discriminant functional analysis.

Keywords: quarrying activities, ground cover plants, environment, discriminant analysis, game reserve

1. Introduction

A quarry is a surface mining-operated area, which produces enormous quantities of gravel, limestone, and other materials for industrial and construction applications [1]. It is a form of land use and part of the local heritage where non-metallic rocks and aggregates are extracted from land [2–4]. Generally, the effects of dust emission from quarries have both micro-spatial and regional dimensions. Air pollution and ground vibration arising from blasting, crushing, and emission of noxious gases have negative impacts on human health and well-being [3]. Several studies have been conducted on the negative impact associated with the environmental effects of quarry activities. One of the biggest negative impacts of quarrying on the environment is the damage to biodiversity, especially the damage caused to plants by pollution resulting from quarrying activities that include necrosis (dead areas on leaf structure), chlorosis (loss or reduction of chlorophyll leading to yellowing of leaf), epinasty (downward curvature of the leaf due to higher rate of growth on the upper surface), and abscission of leaves, that is, premature fall [5–9].

The main focus of this study is to assess the extent of impact of stone-quarrying activities on the ground vegetation, specifically grasses and low-lying herbaceous plants in the Mt. Korok area and its environs within Juba County, Central Equatoria State, South Sudan. The study area constituted a core area of the now-defunct Juba Forest and Game Reserve. The investigation was carried out with the view to provide baseline information on the subject. Moreover, it was hoped that the information obtained would be used to guide policy and management, decision making, and development of mitigation strategies. In addition, the information generated by this study would contribute to the pool of information that many future researchers will use to develop further research concepts and projects.

1.1 Specific objectives were

1. To generate baseline data describing structural composition of grasses and herbal plant communities around the quarrying sites

2. To determine the structural composition of species density, diversity, present ground cover, and frequency distribution at each of the selected site

3. To formulate measures to curb quarrying operations in the area including their impacts on the local environment and to suggest measures that can minimize impacts of quarrying operations

1.2 Quarrying as a human activity

For thousands of years, man has used stone for building, whether it was for monuments, religious buildings, or houses. Early on, man's use of stone and his primitive quarrying would have had little lasting impact on the environment. It was a good material with which to build castles, walls, churches, and important buildings since it was strong and weather resistant [9]. Over the past century, quarrying of building stones and other building material has been on the rise due to increased demand for building material [10]. This has been enhanced by the increased and expansion rate of urbanization locally and internationally [10, 11]. Quarrying is undertaken in different parts of the world, and it impacts the environment and the socio-economic status of the people [12].

The impact on the social economy can be either positive or negative [12–14]. However, the environment is negatively impacted through loss of biodiversity, dust pollution, water pollution, lowering of the water table, soil erosion, and noise pollution [10, 13, 15, 16]. The Victorians, for example, used stone for all their major buildings, and with better transport and new technology, they were able to meet the increasing demands, probably with little thought to their impact on the environment [17, 18]. Today, it is estimated that over 13 million people in about 30 countries across the world are engaged in quarrying, with about 80 million to 100 million people depending on the extractive activities for their livelihood [19].

Quarrying, like many other man's activities, is a process that undergoes different steps, and it involves physically going out into the field and searching for stone; this is then followed by the actual excavation of stone/minerals from the ground [20]. This is achieved in many different ways, depending on what type of stone it is and what you want to take out of the ground. This activity degrades the land after it is quarried, so it is important to study it in order to asses it and to avoid great damage to the environment [18, 20].

As reported in [18, 20, 21], these activities, unfortunately, cause significant impact on the surrounding environment. The extraction process in advanced situations normally depends on heavy machines and explosives, where both processes are normally associated with air pollution, noise pollution, damage to biodiversity, and habitat destruction [14, 22–24]. In addition to water [25], fertile soil is dislocated and interrupted, leaving a big, gaping landscape [26, 27]. The impact can range from scarcely perceptible to highly obtrusive, and the impact can similarly vary widely depending on how quarrying stone was done, the method of quarrying, and the characteristics of the quarry site and its surroundings [28].

One of the biggest negative impacts of quarrying on the environment is the damage to biodiversity, which, essentially, refers to the range of living species, including fish, insects, invertebrates, reptiles, birds, mammals, plants, fungi, and even microorganisms, and quarrying activity has the potential of destroying the habitats and species it supports [7, 23, 24, 29]. Even noise pollution can have a significant impact on some species and affect their successful reproduction, though with careful planning and management, it is possible to minimize the effect on biodiversity, and in fact, quarries can also provide a good opportunity to create new habitats or to restore the existing ones [2, 30].

Dust pollution from quarrying operations both on site and on roads affects the local air quality as well as may lead to serious health. The main potential impacts of dust are visual impacts, coating/soiling of property (including houses, washing bays, and cars), coating of vegetation, contamination of soils, water pollution, change in plant species' composition, loss of sensitive plant species, increased inputs of mineral nutrients, and altered pH balances [24, 28, 31] .

Development requires the utilization of available resources, but very often, it does not check the effects of resource utilization on the environment, [24, 32]. However, quarrying in most developing countries suffers from a number of constraints including a lack of basic knowledge and safety precautions, poor working conditions, low socioeconomic status, lack of clear quarrying legislation, and environmental degradation that call for special attention [4, 5, 33]. On the other hand, quarrying has played a critical role in improving the livelihood of people living in rural areas and town suburbs by creating additional job opportunities and helping to generate additional income [4, 34]. In Africa, East Asia, Southeast Asia, and Latin America, accessibility to natural resources plays a critical role in the livelihood conditions of people; since the formal sectors in developing countries have very little potential in terms of job creation, the informal sector has become an attractive alternative for achieving livelihood needs [4, 35, 36].

1.3 Quarrying activity in South Sudan: past and present

Stones have been quarried in Sudan and South Sudan since the ancient times, at least the beginning of the 14th century and probably long before that. There are multiple records of stone uses in those times; in the middle of the 15th century and after that, stones or quarried stones were used as tools in daily life, for instance, in putting up shelters and many other things, and it was also as much a matter of prestige as of the availability of materials [37]. Most of the stone quarrying was on a small scale and took place close to where the stone was wanted. However, from the recent times, some stone quarries were established due to great demand both locally and further afield in South Sudan for roads and building constructions, especially during the autonomous government in the then Southern Sudan and after the independence of South Sudan in 2011 [38, 39].

2. Materials and methods

2.1 Description of the study area

The area is characterized by a crystalline basement (age: Precambrian) and a hilly and more or less dissected country with well-defined drainage and occasional development of rapid erosion at headwaters [40]. The major physical feature of the reserve is the hilly and rocky outcrops of Mt. Korok itself (also known locally as *Jebel Kujur*, meaning the witch's mountain), which in other parts form very steep and non-negotiable cliffs with trees growing out of their cracks.

Map of the area (see **Figure 1**) encompasses the former Juba Forest and Game Reserve (JFGR). This reserve was established during the British colonial rule in 1939 for the protection of both the unique fauna and flora of the area. It is located to the west of Juba (the capital city of South Sudan) not more than 10 km away from the original Juba town center. The northern boundary of the reserve extends from the mouth of river Lurit to a point on it due north of the summit of Korok Mountain and, from the east, the west bank of *Bahr el Jebel* (The River Nile); the southern boundary is river Dorodo (*Khor Ramla*) from its source to its mouth, and finally, the western boundary is represented by the straight line joining the western limits of the northern and southern boundaries, that is, the town of Juba with its extensions inside the original protected area [41].

The abrupt increase in the population of Juba town as a result of the influx of people from rural areas and the returnees from neighboring countries, following the Peace Agreement of 2005 and the subsequent independence of South Sudan in 2011, has caused a total destruction of JFGR as people scrambled for land to build houses and grow crops. With the subsistent mode of living and their direct dependence on the traditional resources, tree felling, quarrying, charcoal mongering, cultivation, and hunting became rampant and have seriously denuded and deteriorated the area. In the absence of functional laws for protected areas at the time of independence, coupled

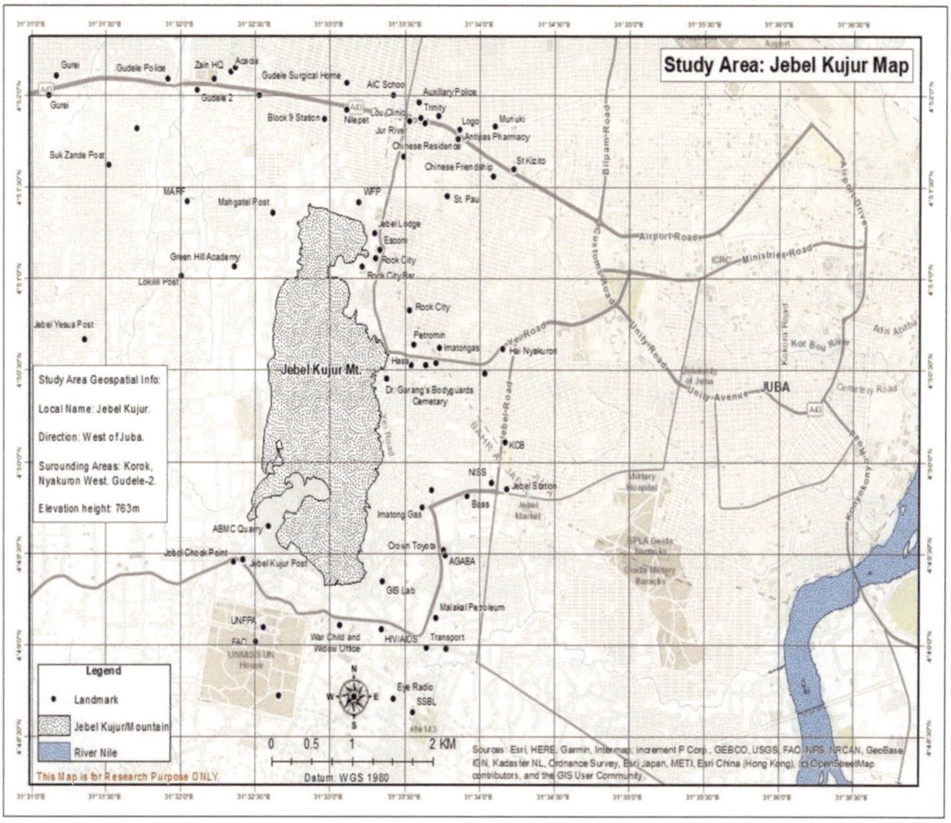

Figure 1.
Map of the study area.

with the ecstasy and euphoria of independence, the Mt. Korok area was gazetted for human habitation the Juba Forest, and Game Reserve was abolished. The degree of environmental degradation that ensued and the disappearance of flora and fauna of the area resulting from human influences have not been properly described or quantified. So, this study is an attempt to describe an aspect of such degradation based on stone-quarrying activities.

The climate is tropical with two distinct wet and dry seasons in a year. The wet season is from April to October and the dry one from November to March. The temperatures are hot all the year round. The minimum temperature ranges from 17 to 25°C during the rainy seasons and the maximum from 25 to 40°C during the dry seasons, with February being the hottest, reaching the maximum of (40°C). Rainfall during the year can reach a total of 1000 mm. Precipitation of more than 1739 mm has been recorded. Rainfall from April to October can reach 99.09 mm per month. Therefore, the climate is characterized as being tropical due to proximity to the equatorial zone within the Central Equatoria state and has savannah vegetation (Directorate of Metrology, South Sudan, 2020).

When the reserve was first established, many different animal species existed therein, including mammals, for example, lion, hyena, leopard, elephant, giraffe, and many kinds of antelopes as well as primates, notably baboon, colabus monkey, vervet monkey, and others. There was also a huge assortment of birds, reptiles, and amphibians. On the other hand, the area was thickly vegetated with woodland savannah species, for example, trees and shrubs such as *Acacia hookii, A. mellifera, A. seiberiana, A. gerrardi, A. senegal, Zizphus spina christi, Z. abyssinica, Piliostigma thoingii, Harisonia abyssinica, Grewia mollis. Tamarindus indica, Celtis integrifolia, Kaya senegalensis, Euhporbia candelabra, Boswelia spp, Xemania Americana Acacia hookii, A. mellifera, A. seiberiana, A. gerrardi, A. Senegal, Zizphus spina christi, Z. abyssinica, Piliostigma thoingii, Harisonia abyssinica, Grewia mollis, T. indica, Celtis integrifolia, Kaya senegalensis, Euhporbia candelabra, Boswelia spp,* and *Xemania Americana* and many grass and herbaceous plant species such as *Hyperrhenia spp, Sprobalus spp, Themeda trianda, Chrysopogon aucheri, Piracharia ruziziensis, Andropogan spp, Cenchrus cililris, Cynodon dactylon, Entropogon macrostachyus, Solanum incanum, Impomea cordofana, Trifolium semifilosum, Cyanthla orthacantha, Dosmodium uncinatum*, and *Indigofera scliperi* to name but a few [41, 42].

2.2 Research design

This study was carried out during the rainy season in the period from June 11 to September 27, 2020. In this study, following initial visits to the area and conducting a pilot survey, three sites were identified and selected to suit the purpose of this study. The first was Site 1, an area that was formerly subjected to stone-quarrying activities but was now abandoned for over seven years (see **Figure 2**).

The second was Site 2, an area where active stone quarrying work was going on at the time of this study (see **Figure 3**) and where it was visibly badly trampled by trucks and pedestrians working on the site and was therefore affected by road pollution including dust, oil spills, and leftover garbage.

The third was Site 3, an area where there had never been any form of stone quarrying activity before and or during the time of this study (see **Figure 4**).

The underlying idea was to sample grass and indeed all low-lying herbaceous plants covering the ground in these selected sites within Korok area by the use of a 1 ×

Figure 2.
Site 1, formerly quarried site now abandoned. Source: field survey, 2020.

Figure 3.
Site 2, actively quarried. Source: field survey, 2020.

Figure 4.
Site 3, an area never subjected to quarrying before. Source: field survey, 2020.

1 m iron frame quadrat, so as to assess the impacts of stone quarrying on the aspects of biodiversity, specifically grass and low-lying plants within those delineated areas.

2.3 Data collection techniques

Impacts of stone quarrying on ground cover vegetation in Korok area were described through the collection of data on abundance, frequency distribution, diversity, and ground cover of grass and herbaceous plants. The objective was to determine how density varied between sites as a consequence of stone-quarrying activities or the absence thereof. A 1 m^2 quadrat was used to estimate the density of grasses and low-lying herbaceous plants in these sites. 100 m^2 of Korok area was delimited, and the number of grass species and other low-lying plant cover within 10 randomly placed quadrats was then counted. The random placing of the quadrats within the area was achieved by the use of a random number table to define the upper left-hand corner of each quadrat. This sampling protocol was repeated at the three chosen sites, and it followed that:

1. Density of grass species was measured by counting the number of individuals of the study species within each quadrat. Only the plants rooted within the quadrat were counted.

2. Visual estimate of percentage cover was made within each quadrat rather than for the whole study area, and later on, these were summarized and extrapolated for the whole area following Domin and Braun-Blanquet scales for visual estimates of cover [43].

3. Frequency of occurrence was calculated as the percentage of the quadrats placed in which the species were observed [44].

4. Grass species and low-lying herbal plants were measured by harvesting the above-ground parts of the plants—which were cut at a certain height from the surface of the substratum, usually at or close to the ground level. A knife, scissors, shears, saw, or chainsaw was used, depending on the grass type. The plant materials were then taken to the laboratory, in bags or sacks when possible, and then sorted into species for purposes of determining diversity indices [43].

2.4 Data analysis

2.4.1 Analyzing community diversity

To determine three important community characteristics:

- Species richness within each site
- Species diversity within each site
- The similarity of grass and herbaceous plant communities between sites

Species richness is simply the tally of different grass species that were identified in a site. Species diversity is a more complex concept. In this work, a standard index called Simpson's Reciprocal was used:

$$D = \sum p_i^2 \tag{1}$$

Where p_i = the fractional abundance of the *i*th species on a site.

Thus, the higher the value, the greater is the diversity. The maximum value is the number of species in the sample, which occurs when all species contain an equal number of individuals. Because this index reflects not only the number of species present but also the relative distribution of individuals among species within a community, it can reflect how "balanced" communities are in terms of how individuals are distributed across species, sometimes referred to as "evenness" [45, 46]. As a result, two communities may have an identical complement of species, and hence species richness, but substantially different diversity measure if individuals in one community are skewed toward a few of the species, whereas individuals are distributed more evenly in the other community.

2.4.2 Analyzing community distinctiveness

Another important perspective in ranking sites is how different the communities are from one another. The simplest available measure of community similarity used here was the Jaccard Coefficient of Community Similarity (CC_J), to contrast community distinctiveness between all the possible pairs of site [27]:

$$CC_J = c/S \tag{2}$$

Where c is the number of species common to both communities and S is the total number of species present in the two communities.

This distinctiveness or similarity index measures the degree to which the species and relative abundance are shared between different grass communities. And the index ranges from 0 (when no species is found in common between communities) to 1 (when all species are found in both communities). In other words, completely similar communities have an index of 1, while completely dissimilar communities have an index of 0. This index was calculated to compare each pair of sites separately, that is, compare Site 1 with Site 2, Site 1 with Site 3, and Site 2 with Site 3 for a total of three comparisons.

2.4.3 Testing for difference of observed median densities

To test if the medians of the observed densities at the three sites were significantly different, a Kruskal-Wallies test was run to compare the medians. And the null hypothesis in this case was that there was no significant difference in the median densities of grass and herbaceous plants species at the three chosen sites.

2.4.4 Calculation of prominence value (PV)

This is a measure of distribution and abundance for each species. This value weighs species abundance by its frequency of occurrence at sample points within each study site [47]. The formula involved is:

$$PV = D \times \sqrt{F} \tag{3}$$

Where PV = number of individuals (D) of each species from all counts in each site multiplied by the square root of frequency of occurrence (F).

Cover is a measure of the area covered by the above-ground parts of plants of a species when viewed directly from above [44]. Because the vegetation may be layered, the cover of all species often sums to more than 100%. In this study, cover was assessed and analyzed according to Domin and Braun-Blanquet scales for visual estimates [43].

2.4.5 Ground cover estimation

Although density and frequency indicate numbers and distribution, they do not indicate size, volume of space occupied, or amount of ground covered or shaded. These characteristics are desirable additional values that contribute materially to an understanding of the importance of a species in a stand, since they are closely related to dominance [48]. Cover can be estimated with some success or may be accurately determined by various devices for measurement and recording. When vegetation is stratified, the cover must be considered in terms of the stratum, zone, or site to which the species belongs. For rapid estimation as well as for analysis of results, there is a distinct advantage at times in using cover classes rather than the specific values; classes of the following number and magnitude are commonly used, covering: (a) less than 5%, (b) 5–25%, (c) 25–50%, (d). 50–75%, and (e) 75–100% [48].

3. Results and discussion

3.1 Plant cover species identified

Principally, the grass family is undoubtedly the most important plant family, and the fifth largest plant family on earth [49]. The largest plant families (in decreasing order) are the Asteraceae (sunflower family), Fabaceae (legumes), Orchidaceae (orchids), and Rubiaceae (gardenia family). **Table 1** shows the grass and other

Family	Species	Stone quarry**		
		Site 1$^{\alpha}$	Site 2$^{\beta}$	Site 3$^{\theta}$
Nyctaginaceae	*Boerhavia diffusa*	+		
Poaceae	*Dactyloctenium aegyptium*	+		
	Urochloa poicoides	+		
	Brachiaria ramose	+	+	
	Eragrostis biflora	+		
	Cymbopogan validus	+		
	Digitaria longiflora	+		
	Urochloa mosambiansis	+		
	Urochloa oligotrich	+		
	Digitaria fermata		+	
	Eragrostis ciliaris		+	+
	Panicum laetum		+	
	Cymbopogan spp 1?*			+
	*Eragrostis pseudoseneranth*a			+
	Eragrostis spp 1?*			+
	*Hyperrhenia spp*1?*			+
	Aristida pallida			+
	Eragrostis spp 2?*			+

Family	Species	Stone quarry**		
		Site 1$^{\alpha}$	Site 2$^{\beta}$	Site 3$^{\theta}$
	*Cymbopogan spp 2?**			+
Acanthaceae	*Monechma ciliatume*	+		+
Graminae	*Cynodon lemfuensis*	+	+	+
	Cynodon dactylon	+	+	+
	Chloris pychothrix		+	
	Chloris virgate		+	
	Chloris gayana		+	+
Anaranthaceae	*Achyranthes aspera*	+	+	+
	Amaranthus graecinzans			+
Asclepiadaceae	*Leptodenia hastate*	+		
Rubiaceae	*Mitracarpus scaber*	+		
Fabaceae	*Crotalaria podocarpa*	+		
Euphrbiaceae	*Euphorbia herta*	+	+	+
Commelinaceae	*Commelina colona*	+		
Pedaliaceae	*Sesamum alatum*	+		
Cyperaceae	*Cyperus rotundus*		+	+
	*Cyperus ssp 1?**			
Mollugnaceae	*Limeum pterocarpum*		+	
Convolvulaceae	*Merremia piñata*			+
Capparaceae	*Cadapa farinose*			+
Salvadoraceae	*Salvadora persica*			+
?	Spp 1?††	+		
	Hibiscus spp. 1?*	+		+
?	Spp 2?††	+	+	+
?	Spp 3?††		+	+
?	Spp 4?††		+	

Species not positively identified.$^{\alpha}$Old, abandoned quarrying site. Site where species existed is marked (+).$^{\beta}$Site with active quarrying operations.††Species not identified but noted.$^{\theta}$Site not quarried before (control).*

Table 1.
A checklist of grass and non-woody herbaceous plants identified at three stone quarrying sites in Korok area.

herbaceous plant species identified during field observations in three locations at Korok area.

In **Table 1**, a total of 44 species of grasses and low-lying non-woody herbaceous plants were identified at three study sites within Korok area. In this study, a list-count quadrat was used as mentioned earlier, that is, a simple tabulation of the species present, where the species were identified and their numbers counted. The species identified were then grouped into 16 families, though the number of families could be more since there were certain species that were not positively identified. The family

Poaceae contained the majority of the species identified:18 (41%). Members of other families were not so common in the area. Notably, some species were conspicuously common and occurred in all the three sites; these included two grass species, namely, *Cynodon lemfuensis* and *Cynodon dactylon* (family Graminae); three of the non-woody herbaceous plant species: *Achyranthes aspera* (family Amaranthaceae) and *Euphorbia hirta* (family Euphorbiaceae); and a third species that was not identified properly and is referred to as *Spp2* (see **Table 1**). Others were observed in any two sites only, whereas there were those that occurred only in Site 1, Site 2, or Site 3.

It is worth noting that of all the species identified in the three study sites altogether, herbaceous plants constituted about 30% and grasses 70%. Site 1 (which was an old abandoned quarrying area) contained 25% of all the plant species (both grasses and herbaceous plants). Usually, herbaceous plants are invaders that tend to recolonize an area that was formally disturbed but now left to rest [50]. Site 2 contained about 14% only of all the grass and herbaceous plant species identified; and these belonged to three families (two grasses: families Poaceae and Graminae and one herbaceous plant: family Mollugnaceae). And the species that were found only in this site included: *Brachiaria ramose*, *Digitaria fernosa*, *Chloris pychothrix*, *Chloris virgate*, and *Limeum pterocarpum*, and they are disturbance tolerant [51]. Site 3 (an area not subjected to stone quarrying before) contained about 23% of all the grass and non-woody herbaceous plant species belonging to four families, three of which were non-woody herbaceous plants (family Convolvulaceae, family Capparceae, and family Salvadoraceae) and one grass family with seven species (family Graminae). Characteristically, the two sites where no stone-quarrying activities existed (Site 1 and Site 3) had a strong presence of and higher densities of non-woody herbaceous plant species.

3.2 Community structure: diversity, species richness, and similarity

In analyzing and discussing the community structure of the local vegetation ground cover, the focus is on three community characteristics, namely:

- Species richness within each site
- Species diversity within each site, and
- The similarity of plant communities between sites

Species richness is simply the tally of different plant species that are observed in a site. Species diversity is a more complex concept. In this study, it was obtained using a standard index, the Simpson Reciprocal Index (see **Table 2**); the higher the value, the greater is the diversity. So, the results showed that Site 1 (an old, abandoned quarry area) had the greatest diversity followed by Site 3 (a non-quarry area), whereas Site 2 (an actively quarried area) had the least diversity. Because this index reflects not only the number of species present but also the relative distribution of individuals among species within a community, it can reflect how 'balanced' communities are in terms of how individuals are distributed across species [27]. As a consequence, two communities may have a more or less identical complement of species, and hence species richness (just as in the case of Site 2 and Site 3) but substantially different diversity measures if individuals in one community are skewed

DOI: http://dx.doi.org/10.5772/intechopen.109707

Species	Site 1^{α}	*pi*	pi^2	Site 2^{β}	*Pi*	pi^2	Site 3^{θ}	*pi*	pi^2
Boerhavia diffusa	12	0.0667	0.0044						
Dactyloctenium aegyptium	6	0.0333	0.0011						
Urochloa poicoides	1	0.0056	0.0000						
Brachiaria ramose	2	0.0111	0.0001	66	0.2833	0.0802			
Eragrostis biflora	9	0.0500	0.0025						
Cymbopogan validus	3	0.0167	0.0003						
Digitaria longiflora	2	0.0111	0.0001						
Urochloa mosambiansis	2	0.0111	0.0001						
Urochloa oligotrich	5	0.0278	0.0008						
Digitaria fermata				92	0.3948	0.1559			
Eragrostis ciliaris				50	0.2146	0.0460	7	0.0292	0.0009
Panicum laetum				7	0.0300	0.0009			
*Cymbopogan spp 1?**							8	0.0333	0.0011
Eragrostis pseudosenerantha							10	0.0417	0.0017
*Eragrostis spp 1?**							5	0.0208	0.0004
*Hyperrhenia spp1?**							25	0.1042	0.0109
Aristida pallida							2	0.0083	0.0001
*Eragrostis spp 2?**							12	0.0500	0.0025
*Cymbopogan spp 2?**							8	0.0333	0.0011
Monechma ciliatume	4	0.0222	0.0005				11	0.0458	0.0021
Cynodon lemfuensis	97	0.5389	0.2904	244	1.0472	1.0966	57	0.2375	0.0564
Cynodon dactylon	2	0.0111	0.0001	42	0.1803	0.0325	15	0.0625	0.0039
Chloris pychothrix				7	0.0300	0.0009			
Chloris virgate				2	0.0086	0.0001			
Chloris guayana				1	0.0043	0.0000	3	0.0125	0.0002
Achyranthes aspera	2	0.0111	0.0001	5	0.0215	0.0005	8	0.0333	0.0011
Amaranthus graecizans							3	0.0125	0.0002
Leptodenia hastate	1	0.0056	0.0000						
Mitracarpus scaber	4	0.0222	0.0005						
Crotalaria podocarpa	9	0.0500	0.0025						
Euphorbia herta	2	0.0111	0.0001	5	0.0215	0.0005	26	0.1083	0.0117
Chloris guayana				1	0.0043	0.0000	3	0.0125	0.0002
Achyranthes aspera	2	0.0111	0.0001	5	0.0215	0.0005	8	0.0333	0.0011
Amaranthus graecizans							3	0.0125	0.0002
Leptodenia hastate	1	0.0056	0.0000						
Mitracarpus scaber	4	0.0222	0.0005						
Crotalaria podocarpa	9	0.0500	0.0025						

Species	Site 1^{α}	*pi*	*pi*^{2}	Site 2^{β}	*Pi*	*pi*^{2}	Site 3^{θ}	*pi*	*pi*^{2}
Euphorbia herta	2	0.0111	0.0001	5	0.0215	0.0005	26	0.1083	0.0117
Commelina colona	2	0.0111	0.0001						
Sesamum alatum	3	0.0167	0.0003						
Cyperus rotundus				35	0.1502	0.0226	126	0.5250	0.2756
Cyperus ssp 1?[*]							36	0.1500	0.0225
Limeum pterocarpum				5	0.0215	0.0005			
Merremia piñata							94	0.3917	0.1534
Cadapa farinose							32	0.1333	0.0178
Salvadora persica							2	0.0083	0.0001
Spp 1?[††]	8	0.0444	0.0020						
Hibiscus spp 1?[*]	3	0.0167	0.0003				2	0.0083	0.0001
Spp 2?[††]	1	0.0056	0.0000	22	0.0944	0.0089	6	0.0250	0.0006
Spp 3?[††]				4	0.0172	0.0003	2	0.0083	0.0001
Spp 4?[††]				8	0.0343	0.0012			
$\sum$	180		0.3065	233		1.4476	240		0.5644
Simpson's Reciprocal Index	= 1/D (and D = $\sum p_i^2$)		3.2628			0.6908			1.7719

[*]*Species not positively identified.*[α]*Old, abandoned quarrying site.*[β]*Site with active quarrying work.*[θ]*Site not quarried before (control).*[††]*Species not identified but noted.*

Table 2.
Comparing grass and other herbaceous plant species diversity, species richness, and community similarity indices among the three stone-quarrying sites in Korok area.

toward a few of the species, whereas individuals are distributed more evenly in the other community.

Diversity is one thing, but distinctiveness is quite another. Thus, another important perspective in ranking sites is how different the communities are from one another [27]. The simplest available measure of community similarity is calculated as shown in **Table 3**; this index ranges from 0 (when no species is found in common between communities) to 1 (when all species are found in both communities). So, as shown in **Table 3**, this index was calculated to compare each pair of sites separately, that is, compare Site 1 with Site 2, Site 1 with Site 3, and Site 2 with Site 3, a total of three comparisons.

It is useful to determine the average similarity of one community within a site to all the others, by averaging the (CC_j) values across each comparison in which a particular site is included. Once the calculations of diversity (species richness and Simpson's Reciprocal Index) as in **Table 2** and distinctiveness (CC_j) as in **Table 3** are made, the primary question of how the three selected sites should be ranked in terms of the extent of stone-quarrying impacts on the local vegetation cover, making an informed decision that requires reconciling the analysis of community structure with concepts of biological diversity as it pertains to diversity and distinctiveness, could be answered.

Pair of sites compared	JCC* = c/S**	Number of common species	Total species in both sites
Site 1 vs. Site 2	0.1875 ≈ 18.8%	6	32
Site 1 vs. Site 3	0.1842 ≈ 18.4%	7	38
Site 2 vs. Site 3	0.3 ≈ 30%	9	30

**For detailed calculation of JCC = c/S, see (Appendix A). **c = number of species common to both communities (being compared) and S = total number of species present in the two communities.*

Table 3.
Using the Jaccard Coefficient of Community Similarity, JCC = c/S [27], the contrast community distinctiveness between all the possible pairs of sites.

The decisions can be based principally on the estimates of species richness, diversity, endemicity (species found at only one site), and community similarity. However, once those decisions are made, it might also be good to look at the spatial arrangement of the selected sites and compare that to the species distributions. This might help in the interpretation of the species distributions and might give useful additional information for ranking the sites according to the damage caused by stone-quarrying operations.

3.3 Frequency distribution

As referenced in [43], under some circumstances, it may not be practicable to make actual counts, but plentifulness may be rapidly estimated according to some scale of abundance such as very rare, rare, infrequent, abundant, and very abundant. Such estimates are particularly useful when several similar stands of uniform composition are to be surveyed within a limited time, as was the case in this study. When there is time for adequate sampling, the determination of actual numbers by counting is of greater value, because it permits the expression of density, which is the abundance by number on a unit-area basis. During the field observations in this study, deliberate efforts were dedicated in order to identify and count plant species so that their density could be determined numerically. Of course, not all species with equal densities are of equal importance in a community, or need to be similarly distributed. It therefore becomes necessary to interpret density values or to specify other characters that, combined with density, serve to complete the picture. One such value is frequency. This value is an expression of the percentage of sample plots in which a species occurs [43]. Thus, frequency becomes a very useful value, when used in combination with density, for then not only the number of individuals but also how widely they are distributed in the site under study is known. Knowledge of these two quantitative characteristics, in combination, is fundamental to an understanding of the community structure.

It should be emphasized that frequency values cannot be compared unless determined with plots of equal size. The larger the plots, the higher is the frequency. In this study the frame quadrat used was of fixed size $1 \times 1\ m^2$. In [44], it is asserted that frequency may conveniently be grouped into classes, for example, A (1–20%), B (21–40%), C (41–60%), D (61–80%), and E (81–100%). The results of this study showed that the total frequency for each individual species identified and counted was very low, and if the above frequency classes were to be used, then almost all species would have their frequency in the first class, that is, A (1–20%). And this does not tell much and is therefore meaningless. So, instead of expressing frequencies in terms of percentages, it was thought better to express them in terms of prominence values (see **Table 4**), which is a measure of both abundance and distribution [47]. And in **Table 4**, six species stand out very clearly as being the most abundant and widely distributed all over the three study sites and/or at least in two sites; they are in order of magnitude from the highest to the lowest: *Cynodon lemfuensis*, *Cyperus rotundance*, *Brachiaria ramosa*, *Merremia piñata*, *C. dactylon*, and *Digitaria fernata*. Two other species, namely, *Euphorbia herta* and another that was not positively identified but

Species	Site 1$^{\alpha}$	Site 2$^{\beta}$	Site 3$^{\theta}$	Total counted (D)	Total frequency (F)	Prominence value (PV)
Boerhavia diffusa	12			12	8	33.94
Dactyloctenium aegyptium	6			6	2	8.48
Urochloa poicoides	1			1	1	1
Brachiaria ramose	2	66		68	7	179.91
Eragrostis biflora	9			9	2	12.73
Cymbapogan validus	3			3	2	4.24
Digitaria longiflora	2			2	2	2.83
Urochloa mosambiansis	2			2	1	2
Urochloa oligotrich	5			5	1	5
Digitaria fermata		92		92	2	130.11
Erograstis ciliaris	7	50	7	57	4	114
Panicum laetum		7		7	2	9.9
*Cymbopogan spp 1**			8	8	2	11.31
Erograstic pseudoseneranthe			10	10	1	10
*Erograstis spp 1**			5	5	1	5
*Hyperrhenia 1?**			25	25	1	25
Aristida pallida			2	2	1	2
*Eragrostis spp 2?**			12	12	1	12
*Cymbopogan spp 2?**			8	8	1	8
Monechma ciliatume	4		11	15	3	25.98
Cynodon lemfuensis	97	244	57	398	16	1592
Cynodon dactylon	2	42	15	59	5	131.93
Chloris pychothrix		7		7	1	7
Chloris virgate		2		2	1	2
Chloris guayana		1	3	4	2	5.66
Achyranthes aspera	2	5	8	15	3	25.98
Amaranthus graecizans			3	3	1	3
Leptodenia hastate	1			1	1	1
Mitracarpus scaber	4			4	2	5.66

Species	Site 1[α]	Site 2[β]	Site 3[θ]	Total counted (D)	Total frequency (F)	Prominence value (PV)
Crotalaria podocarpa	9			9	3	15.59
Euphorbia herta	2	5	26	33	3	57.16
Commelina colona	2			2	1	2
Sesamum alatum	3			3	1	3
Cyperus rotundus	2	35	126	163	5	364.48
*Cyperus ssp 1?**			36	36	1	36
Limeum pterocarpum		5		5	1	5
Merremia piñata			94	94	3	162.81
Cadapa farinose			32	32	1	32
Salvadora persica			2	2	1	2
Spp 1?[††]	8			8	3	13.86
Hibiscus spp. 1?*	3		2	5	2	7.07
Spp 2?[††]	1	22	6	29	8	82.02
Spp 3?[††]		4	2	6	2	8.48
Spp 4?[††]		8		8	2	11.31

[*]Species not positively identified.[α]Old, abandoned quarrying site.[β]Site with active quarrying work.[θ]Site not quarried before (control).[††]Species not identified but noted.

Table 4.
Determination of prominence value (PV) for all plant species identified in the three selected study sites within Korok area.

referred to as *Spp 2*, were observably and moderately common in abundance and frequency within the study sites.

3.4 Comparing median densities of grasses and non-Woody herbaceous plants

A non-parametric test, the Kruskal-Wallies test, was used to calculate the statistic (K) of the median densities of grasses and non-woody herbaceous plant species in the three selected sites. The null hypothesis was that there is no significant difference in the median densities of grasses and herbaceous plant species in the three sites. But it was found that $K = 21.76$. Comparing this calculated statistic with the tabulated distribution of χ^2 (Appendix C) in [52] at 2 degrees of freedom (df), the calculated value of K far exceeds the tabulated critical values of 5.99 and 9.21 at both $p = 0.05$ and $p = 0.01$, respectively. We therefore reject the null Hypothesis and conclude that there is a highly significant difference between the median densities of the grasses and the non-woody herbaceous plant species in the three sites ($K = 21.76$, $p < 0.01$, Kruskal-Wallies test) (**Table 5**).

3.5 Ground cover estimation

Using the approach mentioned in Section 2.4.5 above, results of this study revealed that Site 3 (a no-quarry area) had the highest cover followed by Site 1 (a formerly quarried area but now abandoned), whereas Site 2 (an area with intense quarrying

Site 1 (old, abandoned quarrying area)	Site 2 (area with active quarrying work going on)	Site 3 (area not quarried before – control)
2 (3)	42 (19)	30 (14)
6 (4)	36 (16)	50 (20)
1 (1.5)	148 (30)	28 (17)
1 (1.5)	84 (25)	51 (21)
25 (7.5)	85 (26)	22 (6)
27 (10)	117 (28)	77 (23)
28 (17)	40 (18)	20 (5)
26 (9)	81 (24)	124 (29)
32 (15)	66 (22)	87 (27)
25 (7.5)	39 (17)	29 (13)

N.B: for detailed calculation, see (Appendix B).

Table 5.
Median densities of grass and herbaceous plant species and ranked scores (in brackets) at three study sites in Korok area.

activities at the time of this study) had the least cover (**Figure 5**). It can therefore be concluded that quarrying operations in Site 2 to a great extent are responsible for the removal of the low-lying vegetation cover in the area.

3.6 Effects of stone quarrying on relative density and abundance of grass and non-Woody plants in these three sites

By comparing plant relative density at Site 1 and Site 2 (**Figure 5**), it is clear that plants at Site 1 (which is an old abandoned quarrying area) have a higher relative

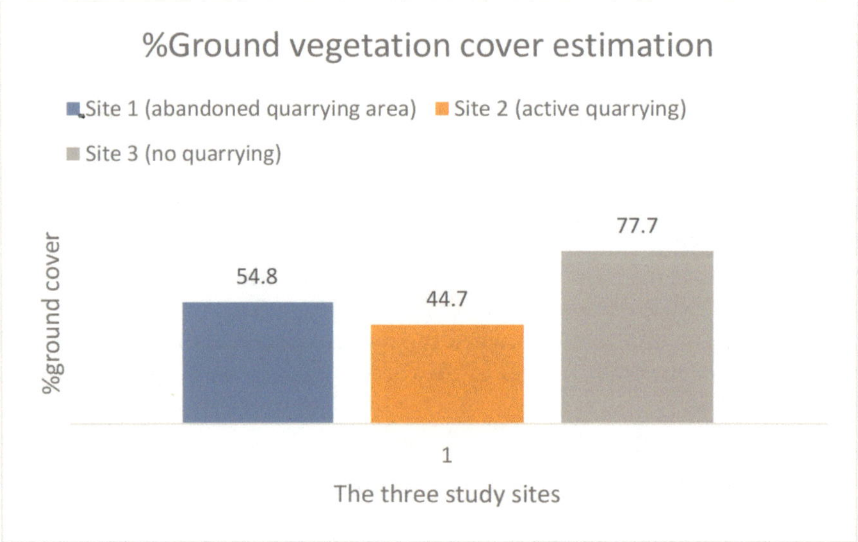

Figure 5.
Proportion of ground covered with grass and herbal plant species in each of the three study sites.

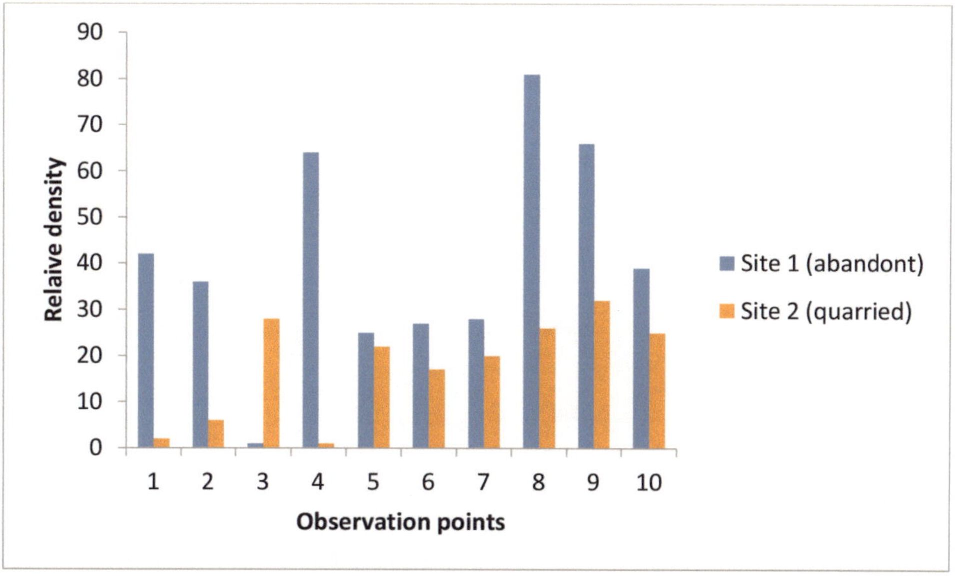

Figure 6.
Relative density and abundance of grass and herbaceous plant species in two study sites: Site 1 (abandoned quarrying area) and Site 2 (area under quarrying activities) within Korok area. (NB. Observation points are the ten sampling points in each site).

density than those at Site 2 (an area with stone-quarrying activities still going on), suggesting that stone-quarrying operations have an impact on the local ground cover (**Figure 6**).

Similarly, upon contrasting relative density of plants at Site 2 with that of Site 3 (an area with no stone-quarrying activities), the results showed that plants at Site 3 had a higher relative density than those at Site 2 (**Figure 7**). In both of these comparisons, results are consistent with the findings of previous researchers. For example, it was well documented how stone-quarrying can affect vegetation cover, which represents the main component of the ecosystem, hence the absence of balance in the volume of oxygen and carbon dioxide through photosynthetic activities [7, 25]. Likewise, it was noted that stone quarrying can indirectly affect local vegetation cover through its negative impact on the soil and pollute both surface and ground water [24, 53, 54]. In the same vein, [26] reported that stone quarrying has resulted in changes in soil properties such that soil in and around the quarrying area (0–1 km) was found to be alkaline (pH 11.2–11.7), and this was attributed to the high concentrations of hydroxyl, carbonate, and bicarbonate present in the minerals of mined materials. In addition to the physical removal of ground cover by tools used in mining and stone-cutting industries, the physiological mechanisms behind plant damage could be attributed to one or a combination of the following factors: dusts might cover the leaves with a white layer, thereby decreasing the total chlorophyll cells exposed to light and thus reducing the total photosynthetic activity [53].

In **Figure 8**, the two sites with apparently no stone-quarrying activities, the relative density of plants therein are both high. This suggests that when an area is not disturbed through human activities such as stone quarrying, the grasses and other non-woody herbaceous plants have the potential to reclaim the area and to establish themselves. And their presence also prevents soil erosion that results from the removal of ground cover by stone-quarrying operations. Plants are exceptionally

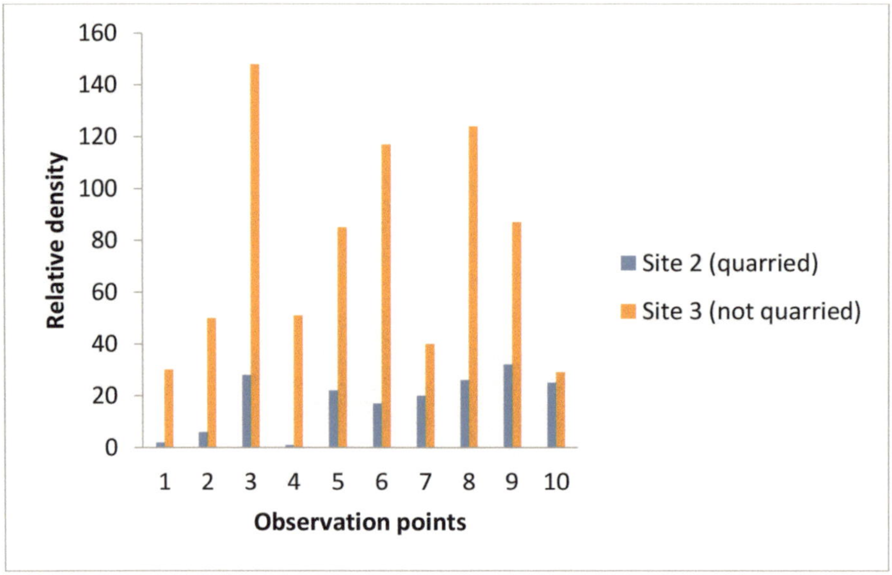

Figure 7.
Relative density and abundance of grass and forb-like plant species in two study sites: Site 2 (area under quarrying) and Site 3 (area with no quarrying activity) within Korok area.

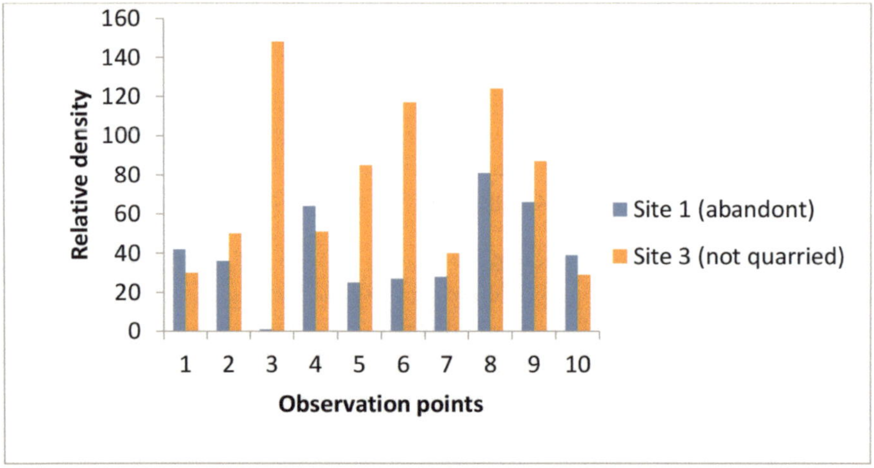

Figure 8.
Relative density and abundance of grass and herbaceous plant species in two study sites: Site 1 (abandoned quarrying area) and Site 3 (area with no quarrying activities) within Korok area.

effective in protecting the soil against the agents of erosion like water, wind, and sun. Grasses are known for being particularly effective in combating soil erosion [49]. Their growth points are very close to the ground level, and they often form stolons and/or rhizomes, which are good at stabilizing the soil [55]. If not disturbed by human activities, plant communities in the area will undergo the natural process of plant succession, which is the progressive succession of plant communities. When a disturbance like stone quarrying and the accompanying excavations takes place in an area, the area is recolonized by new, better-adapted plant communities [49]. The two sites

(1 and 3) share many species between them that are not found in the disturbed Site 2, which also suggests that the plant communities in them might be at a different level of plant succession, hence proving that stone-quarrying work is impacting the local vegetation cover negatively.

The relative densities of plants in the three sites are compared in **Figure 9**, and the difference is glaringly clear. Plants in the area with no stone-quarrying operations (Site 3) have the highest relative density, followed by those in the formerly disturbed but now abandoned area (Site 1), and, lastly, Site 2 (quarried area) plants with the least relative density.

However, despite this apparent difference in the densities of grass cover, the result of a Pearson's Moment Correlation Coefficient test (r) run to test the null Hypothesis that there was no significant difference in impacts due to quarrying activities *per se* in respect to relative plant densities between Site 2 and Site 3, a weak correlation of (r) equals 0.392 was obtained. The (r) value is far lower than the tabulated critical value of 0.632 for degrees of freedom 8 at $p = 0.05$. Therefore, the H_0 is accepted, and it was concluded that the positive correlation was statistically not significant. This suggested that although there was a positive weak correlation between the extent of stone-quarrying activities and relative density of plant constituting the local ground cover in Site 2 vs. Site 3, stone quarrying might not be the only factor affecting the densities of ground cover vegetation, notably grasses and low-lying herbaceous plants; hence, the impact is not significant. To establish this assumption, a coefficient of determination (r^2) was calculated to indicate how much other factors, besides stone-quarrying activities, influenced plant densities, distribution, frequencies, and cover. The coefficient of determination was found to be $r^2 = 0.154$; in other words, only about 15% of the impacts resulted from stone-quarrying activities in the selected study sides, meaning over 85% of variation in ground cover vegetation density was not accounted for by stone quarrying alone. It is probable that other factors such as bush burning, small-scale farming in the area, and construction work like building houses, roads, and so on could be major contributors impacting plant cover and density in the area. This must be investigated in future studies along this same theme. Unfortunately,

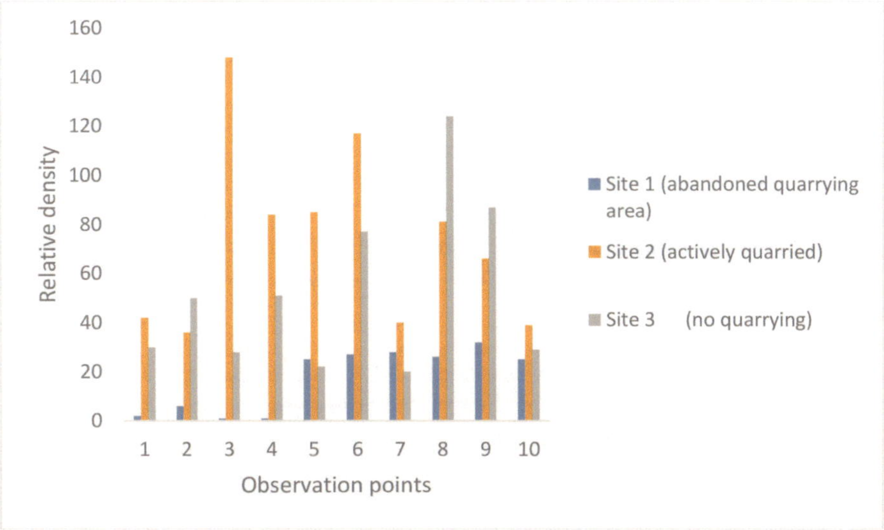

Figure 9.
Relative density and abundance of grass and forb-like plant species in the three study sites within Korok area.

given the short study period and limited finance and logistics, the kind of data collected could not be subjected to discriminant analysis technique so as to effectively isolate the negative effects of stone quarrying from other likely factors. Discriminant analysis is a powerful classification technique to discriminate the assigned observations to predefined groups [56].

Quarrying is a sensitive and a complex issue. On the one hand, quarrying supplies raw materials to meet many of the societies 'needs, creates employment, and contributes to the local economy [12, 57]. This part of South Sudan, that is, the Korok area, has always experienced and is still experiencing extensive quarrying activities. The residents of the area living adjacent to quarrying sites are exposed to disturbances and environmental impacts of stone quarrying: the constant traffic of heavy dumpers and the movements of lorries to and from the sites cause noise and the result of rock blasting generate dust, smoke, and fumes as well as suspended particles, all of which pollute the area. As a consequence, people who live near the quarrying areas are likely to develop respiratory diseases. Observably, in the area, quarrying activities also produce a growing number of abandoned quarry pits that are quickly filled-up with water in the rainy season and become suitable habitats for swarms of mosquitoes and freshwater snails that in turn act as intermediate hosts for *Schistosoma* and *Haematobium* that may eventually contribute to the prevalence of bilharzia, urinary problems, and malaria in people. Apart from these, land degradation and other negative impacts of stone quarrying include swamp creation, deterioration of underground water, and erosion of soil; moreover, quarrying activities such as excavation, digging, blasting, and clearing the land have direct effects on biodiversity.

4. Conclusion and recommendations

4.1 Conclusion

Quarrying and stone-cutting activities affect the general environment including destruction and removal of local vegetation cover, particularly grass and the low-lying non-woody herbaceous plants. The physical crushing, excavation, and removal of rocks produce high concentrations of particulate matter (dust), fumes, smoke, and other gaseous substances, which negatively affect vegetation in the vicinity of quarrying areas. Also, the stone-quarrying activities affect ground cover plants indirectly by affecting soil and water, which are vital resources for vegetation cover, thus exacerbating the problem. However, in the Mt. Korok area, stone-quarrying activities are not the only cause of ground cover vegetation destruction; there are other factors that equally affect the local vegetation cover, but they are yet to be determined.

4.2 Recommendations

1. This study has been limited in its extent as it was directed only toward the impact of stone-quarrying activities on the ground cover vegetation, mainly grasses and low-lying herbaceous plants. Future investigations of the same should be expanded to include the impacts on:

    a. Biodiversity in general

    b. Land environment (landscape forms, land use types)

c. Air environment (dust pollution, noise pollution)

d. Water environment (pollution of surface water and groundwater)

e. Livelihood of the communities living within or adjacent to quarrying sites in terms of their health, socio-economic status, and so on.

2. Future studies on the effect of quarrying on vegetation cover should properly isolate its negative effects on vegetation cover from the effects of other factors such as bush burning; construction works, for example, house and road building; as well as farming by using discriminant analysis, which is a powerful classification technique to discriminate the assigned observations to predefined groups.

3. Environmental impact assessment should be conducted prior to any stone-quarrying plans and should be made in compliance with the state and/or national development plan of the area.

4. Awareness campaigns should be conducted at various levels about the impacts of stone quarrying on vegetation and biodiversity in general in Korok area.

Acknowledgements

We are grateful to the School of Natural Resources and Environmental Studies, University of Juba, for supporting this study by allowing us to use the school's field tools and laboratory facility. We also thank the local government authority and the community leaders in the Mt. Korok area for granting us the permission to carry out the study in their area. Many other people also assisted us in the course of this study; to all of them, we say thank you for the support they gave us.

A. Appendix

Calculating community similarity by using the simplest measure, which is the Jaccard coefficient of community similarity to contrast community distinctiveness between all possible pairs of sites.

$$\text{Jaccard Coefficient of Community Similarity, CCJ} = c/S.$$

Where c = is the number of species common to both communities (being compared) and S = is the total number of species present in the two communities.

So, Site 1 vs. Site 2: is c/S = 0.1875, i.e., 18. 8%. 6 species are common, and total of both sites (22 + 16) − 6 = 32.

Site 1 vs. Site 3: is c/S = 0.184210526, i.e., 18. 4%. 7 species are common, and total of both sites is (22 + 23)−7 = 38. Site 2 vs. Site 3: is c/S = 0.3, i.e., 30%. 9 species are common, and total of both sites is (16 + 23) − 9 = 30.

B. Appendix

Calculating median densities of grass and forb-like plant species at the three sites in Korok area number of individual species observed (rank scores in brackets).

Site 1 (old, abandoned quarrying area)	Site 2 (area with active quarrying work going on)	Site 3 (area not quarried before – control)
2 (3)	42 (19)	30 (14)
6 (4)	36 (16)	50 (20)
1 (1.5)	148 (30)	28 (17)
1 (1.5)	84 (25)	51 (21)
25 (7.5)	85 (26)	22 (06)
27 (10)	117 (28)	77 (23)
28 (17)	40 (18)	20 (05)
26 (09)	81 (24)	124 (29)
32 (15)	66 (22)	87 (27)
25 (7.5)	39 (17)	29 (13)

n = 10	10	10	N = 30
R = 76	225	175	
R^2 = 5776	50625	30625	
R^2/n = 577.6	5062.5	3062.5	
$\sum(R^2/n)$ = 8702.6			

$$K = \left[\sum(\mathrm{R}^2/\mathrm{n}) \times 12/\mathrm{N}/\mathrm{N}+1\right] - 3(\mathrm{N}+1)$$
$$= [8702.6 \times 12/30/(31) - 3(31)$$
$$K = 21.76$$

We compare K with the tabulated distribution of χ^2 (Appendix C) [52]. The degree of freedom is the number of samples less one (in this case 3–1 = 2). At 2 df, our calculated value far exceeds the tabulated critical values of 5.99 and 9.21 at both p = 0.05 and p = 0.01, respectively. We therefore reject the null Hypothesis and conclude that there is a highly significant difference between the median densities of the grass and herbaceous plant species in the three sites (K = 21.76, $p < 0.01$, Kruskal-Wallies test).

C. Appendix

Calculating the Product Moment Correlation Coefficient, r, for Study Site 2 vs. Study Site 3 at Korok area.

Site 2* (x)	Site 3* (y)	x^2	y^2	xy
42	30	1764	900	1260
36	50	1296	2500	1800
148	28	21,025	784	4144

Site 2* (x)	Site 3* (y)	x^2	y^2	xy
84	51	7056	2601	4284
85	22	7225	484	1870
117	77	13,689	539	9009
40	20	1600	400	800
81	124	6561	15,376	10,044
66	87	4356	7569	5742
39	29	1521	841	1131
$\sum 738$	$\sum 518$	$\sum 58{,}793$	$\sum 31{,}994$	$\sum 40{,}084$

$(\sum x)^2$ = 544,644.
$(\sum y)^2$ = 268,324.

The Product Moment Correlation Coefficient (r) is calculated as follows:

$$
\begin{aligned}
r &= \frac{n\sum xy - \sum x \sum y}{\sqrt{\left[n\sum x^2 - (\sum x)^2\right]\left[n\sum y^2 - (\sum y)^2\right]}} \\
&= \frac{10 \times 40084 - 738 \times 518}{\sqrt{[10 \times 58793 - 544644][10 \times 31994 - 268324]}} \\
&= \frac{400840 - 382284}{\sqrt{(43286)(51616)}} \\
&= \frac{18556}{\sqrt{2234250176}} \\
&= \frac{18556}{47267.8} \\
&= 0.392
\end{aligned}
$$

The correlation coefficient appears weak and positive. Checking the significance of this positive correlation from appendix 5 in [52], we find that the value 0.392 far below the tabulated critical value of 0.632 for degree of freedom (n – 2), i.e., 10–2 = 8 at p = 0.05. We accept the H_0 and conclude that the weak positive correlation is statistically not significant.

D. Appendix

Calculation of Coefficient of Determination (r^2).

This is done by squaring the Product Moment Correlation Coefficient, r, [52], i.e., 0.392.

Thus, $r^2 = 0.392^2 = 0.153667$.

Author details

Pasquale Tiberio Moilinga[1*] and Makuac Robert Athian[2]

1 Department of Wildlife Science, School of Natural Resources and Environmental Studies, University of Juba, Juba, South Sudan

2 Department of Geography, School of Arts and Humanities, University of Juba, Juba, South Sudan

*Address all correspondence to: worimoilinga@gmail.com

References

[1] Duan W, HaiRen H, Fu S, Wang J, Yang L, Zhang J. Natural recovery of different areas of a deserted quarry in South China. Journal of Environmental Sciences. 2008;**20**(4):476-481

[2] Ukpong EC. Environmental impact of aggregate mining by crush rock Industries in Akamkpa Local Government Area of Cross River state. Nigerian Journal of Technology. 2014;**31**: 116-127

[3] Nartey VK, Nanor JN, Klake RK. Effects of quarry activities on some selected communities in the lower Manyakrobo District of the Eastern Region of Ghana. Atmospheric and Climate Sciences. 2012;**2**:362-372

[4] Nyapala OA, Kamwele H. Socio economic impact assessment of stone quarrying in Thika municipality: A case study of Nanasi Area Block 14 (2010-2011). In: 4th World Conference on Applied Sciences, Engineering and Technology. 24–26 October. Japan: Kumamoto University; 2015

[5] Wang Q, Shi J, Chen G, Xue L. Environmental effects induced by human activities in arid Shiyang River basisn, Gansu Province, Northwest China. Environmental Geology. 2002;**43** (1–2):219-227

[6] Khater C, Martin A, Maillet J. Spontaneous vegetation dynamics and restoration prospects for limestone quarries in Lebanon. Applied Vegetation Science. 2003;**2**:199-204

[7] Anand PB. Waste management in Madras revisited. Environmental Urbanization. 2006;**11**(20):161-176

[8] Jomaa I, Auda Y, Abi Saleh B, Hamze M, Safi S. Landscape spatial dynamics over 38 years under natural and anthropogenic pressures in Mount Lebanon. Landscape and Urban Planning. 2008;**87**:67-75

[9] Xue YG, Teng DB, Li SC, Su MX. Study on Environmental-Geological Problem and Ecosystem Re-Establishment Countermeasures about Quarry and Damaged Mountain in Peri-Urban. In: 4th International Conference on Bioinformatics and Biomedical Engineering (i CBBE 2010). China; 2010, 2010. pp. 3717-3721

[10] Dong-dong Z, Yu-shan S, Le L. Study on sustainable landscape design of abandoned quarries. Procedia Earth and Planetary Science. 2009;**1**(1):1107-1113

[11] Afeni TB. Assessment of the Socio-Economic Impacts of Quarrying and Processing of Limestone at Obajana, Nigeria. School of Mining Engineering, University of The Witwatersrand, Johannesburg, South Africa. European Journal for Social Sciences. 2008

[12] Lad RJ, Samant JS. Environmental and socil impacts of stone quarrying: A case study of Kolhapur district. International Journal of Current Research. 2014;**6**(3):5664-5669

[13] Chatterjee N. The basalt stone quarries of eastern India; the basalt stone quarries of eastern India. International Journal of Environmental Studies. 2010; **67**(3):439-457

[14] Sati VP. Socio-economic and environmental impacts of stone Mining in Shivpuri District, Madhya Predesh, India. Journal of Scientific Research and Reports. 2015;**4**:47-54

[15] Edit A, Nganje AJ, Ekwere AS, Ukpong AJ. Groundwater chemistry and

quality of Nigeria: A status review. African Journal of Environmental Sciences and Technology. 2011;**5**(13): 1152-1169

[16] Kaliampakos DC, Mavrikos AA. Introducing a new aspect in marble quarry rehabilitation in Greece. Environmental Geology. 2006;**50**: 353-359

[17] Bloodworth A, Scott J, Mcevoyc FM. A British Geological Survey. Keyworth: Kingsley Dunham Centre; 2009

[18] USMI. Stone and Marble in Palestine, Developing a Strategy for the Future. Union of Stone and Marble Industry; 2011

[19] Asante F, Abass K, Afriyie K. Stone quarrying and livelihood transformation in Peri-urban Kumasi. Research on Humanities and Social Sciences. 2014; **4**(13):93-108

[20] Gamal El-Dine HM, Sadek RR, Zayet HH, Refaat TM. Respiratory problems among workers exposed to quarries dusts in El-Mina governorate. El Mina Medical Bulletin. 2009;**20**(2):360

[21] Okafor FC. Rural Development and the Environmental Degradation versus Protection. In: Sada PO, Odemerho T, editors. Environmental Issues and Management in Nigerian Development. 2006. pp. 150-163

[22] Abate Z. Impacts of Stone Quarrying on Environment and Livelihood of Local Community in Addis Ababa Peri-Urban Areas: The Case of Hana Mariam Cobble Stone Quarry Site. MSc Thesis. Addis Ababa, Ethiopia: Addis Ababa University; 2016

[23] Maboguine AL. The Challenges of Mobility within Nigeria's Energing Megacities. In: Keynote Address Delivered at the Maiden Annual National Conference on Public Transport. Ikeja; 2008. pp. 6-8

[24] Lameed GA, Ayodele AE. Effects of quarrying activity on biodiversity: Case Study of Ogbere site, Ogun State, Nigeria. African Journal of Environmental Science and Technology. 2010;**4**(11):740-750

[25] Osha OL. Information Booklet on Industrial Hygiene. Revised ed. Washington, USA: U.S. Department of Labor OSHA/OICA Publications, Occupational Safety and Health Administration; 2006. pp. 23-35

[26] Haritash AK, Baskar R, Shama N, Paliwal S. Impact of slate quarrying on soil properties in semi-arid Mahendragarh in India. Environmental Geology. 2007;**51**:1439-1445

[27] Gibbs JP, Hunter ML, Jr; and Sterling, E. J. Problem-Solving in Conservation Biology and Wildlife Management: Exercise for Class, Field, and Laboratory. 2nd ed. Oxford: Blackwell Publishing; 2008

[28] Angela AO, Mba C, Uloma JU. Effects of quarrying activities on local vegetation cover in Ebonyi State, Nigeria. International Journal of Science and Research Methodology. 2017;**6**(2): 35-50

[29] Nyakeniga CA. An Assessment of Environmental Impacts of Stone Quarrying Activities in Nyambera location, Kisii County. Kenyatta University; 2009

[30] Endalew A, Tasew E, Tolahun S. Environment and social impacts of stone quarrying: South Western Ethiopia, in case of Bahir Dar Zuria Wereda Zenzelma Kebele. International Journal

of Research in Environmental Science (IJRES). 2019;**5**(2):29-38

[31] Tanko A. Environmental Concerns, Assessment and Protection Procedures for Nigeria's Oil Industry. Nigeria: Center for Development Studies and the School of Geography; 2007. p. 1

[32] Oke SO, Ibanesebhur G. Impact of limestone quarrying on vegetation and landform of Ewekoro Cement, Ewekoro Local Government Area, Ogun State, Nigeria. Journal of Botany. 2010;**23**(2): 301-368

[33] Foddy G, Wallen E, Agarwall D. Resolving Social Dilemma. Philadelphia, P A: Psychology Press; 1999

[34] Mouflis GD, Gitas LZ, Lliadou S, Mitri GH. Assessment of the visual impact of marble quarry expansion (1984–2000) on the landscape of Thasos Island, NE Greece. Landscape and Urban Planning. 2008;**86**:92-102

[35] Aigbedion IN. Environmental pollution in the Niger-Delta, Nigeria. Inter Disciplinary Journal of Environment. 2005;**3**(4):205-211

[36] Adekoya JA. Environmental effect of solid minerals mining. Journal of Physical Science. 2003;**4**(2):625-640

[37] Bloxam E, Helda T. The industrial landscape of the northern Faiyum Desert as a world heritage site: Modelling the outstanding universal value of third millennium BC stone quarrying in Egypt' in world archaeology 393. The Archaeology of World Heritage. 2007; **2007**:305-323

[38] Dirks PH, Blenkisop TG, Jelsma HA. The Geological Evolution of Africa. Paris: United Nations Educational, Scientific and Cultural Organization (UNESCO); 2002

[39] Cordaid. Mining in South Sudan Opportunities and Risks for Local Communities. Baseline Assessment of Small-Scale and Artisanal Gold Mining in Central and Eastern States, South Sudan. Juba: Catholic Organisation for Relief and Development Aid (cordaid); 2016

[40] Mefit SPA. Regional Development Plan, First Phase B. Rome: Mefit SPA; 1977

[41] Hillman JG. Wildlife Information Booklet, Department of Wildlife Management. USA: N.Y.Z.S; 1982

[42] Noordjwik M. Ecology Text Book for the Sudan. Sudan: Khartoum University Press; 1984

[43] Sutherland WJ. Ecological Census Techniques: A Handbook. United Kingdom: Cobridge University Press; 1996

[44] Grieg-Smith P. Quantitative Plant Ecology. 3rd ed. UK: Blackwell Scientific Publications. Oxford; 1983

[45] Begon M, Herper JL, Townsend CR. Ecology: Individuals, Populations and Communities. 2nd ed. UK: Blackwell Scientific Publications; 1990. p. 945

[46] Gibbs JP, Harrison IJ, Griffiths J. What is biology? Spiders as examples of the biodiversity concept. In: Gibbs JP, Harrison IJ, Jennifer G, editors. Problem-Solving in Conservation Biology and Wildlife Management. 2nd ed. UK, Australia: Blackwell Publishers; 2008

[47] Steel RGD, Torrie JH. Principles and Procedures of Statistics, with Special Reference to the Biological Science. NY. Toronto, London: McGrow-Will Book Company, Inc; 1960

[48] Wheater CP, Bell JR, Cook PA. Practical Field Ecology: A Project Guide.

1st ed. UK: JohnWiley & Sons, Publications; 2011

[49] Van Oudtshoorn F. Guide to Grasses of South Africa. Pretoria, South Africa: Briza Publications; 1992

[50] Smith RL. Ecology and Field Biology. 4th ed. New York, USA: Harper Collins Publishers, Inc.; 1990

[51] Begon M, Mortimer M, Thompson DJ. Population Ecology a Unified Study of Animals and Plants. 3rd ed. UK: Blackwell Science Ltd; 1996

[52] Fowler J, Cohen L, Jarvis P. Practical Statistics for Field Biology. 2nd ed. West Sussex, England: John Wiley & Sons Ltd; 1998

[53] Missanjo E, Kamanga-Thole G, Mtambo C, Chisinga O. Evaluation of natural regeneration and tree species diversity in Miombo woodlands in Malawi. Journal of Biodiversity Management and Forestry. 2014;**3**(3):4

[54] Prajapati SK, Tripathi BD. Anticipated performance index of some tree species considered f1349.or green belt development in and around an urban area: A case study of Varanasi city; Indian. Journal of Environmental Management. 2008;**88**:1343-1349

[55] Henderson PA. Practical Methods in Ecology. Cornwall, UK: MPC Books Ltd, Bodmin; 2003

[56] George CJF. Discriminant Analysis. A Powerful Classification Technique in Data Mining. Reno: University of Nevada; 2014

[57] Hardin G. The tragedy of the common. Science. 1968;**162**:1243-1248

Chapter 4

Perspective Chapter: A Perspective on the Resettlement of Maasai Communities from the Ngorongoro Landscape in Tanzania

Franco P. Mbise

Abstract

Tanzania is endowed with an abundance of natural resources, and the country has set aside more than 30% of its land for conservation. The Ngorongoro district is where both the Ngorongoro Conservation Area (NCA) and the Loliondo Game Controlled Area (LGCA) are located. Sometimes, it is confusing when talking about Ngorongoro as a region and the NCA as a conservation area. The NCA is an example of a diverse landscape in which humans (the Maasai tribe) and their livestock coexist with wildlife, but crop farming is prohibited. According to current projections, the Maasai population at NCA is approximately 120,000 people. And there are approximately 45,200 livestock. The increase in people is proportional to the increase in livestock, as the Maasai who live in the Ngorongoro landscape rely solely on animal husbandry as their main source of income. In the NCA, human and livestock population growth have a significant impact on wildlife and their habitat. To alleviate the pressure on the landscape, the Tanzanian government took the initiative to seek Maasai willing to relocate to another area, in this case, Handeni district in Tanga region.

Keywords: landscape conservation, Maasai, quality of life, resettlement, Tanzania

1. Introduction

The International Union for Conservation of Nature (IUCN) has defined six Protected Areas Management Categories that describe protected areas (PAs) and their management planning through the World Commission on Protected Areas (WCPA). This includes sustainable use areas set aside to protect important ecological processes, species, and ecosystems while also providing cultural, scientific, educational, recreational, and visitor opportunities and strictly protected areas set aside to protect biodiversity and only used for scientific research and monitoring [1].

The Yellowstone National Park in the United States, Kruger National Park in South Africa, and Serengeti National Park in Tanzania are all well-known parks with great conservation histories. Protected areas around the world have become the only way to protect species and their habitats. As the human population grows, these protected areas are increasingly encroached upon by locals, degrading and destroying these pristine wildlife habitats. Recently, the focus of wildlife conservation has shifted to single species, particularly endangered species. The main concern is to protect the entire ecosystem, as this will protect all species. True, threatened and endangered species require extra care in conservation. This initiative should be handled with extreme caution without losing sight of the importance of conserving the entire ecosystem, which in turn conserves all species. Nowadays, conservationists understand that without local communities, the paradigm of sustainable conservation will fade away. Communities living within or adjacent to protected areas play a critical role in achieving this goal. When people in a community care about conservation, it costs the government less to protect protected areas [2–5]. A major issue arises when the two actors are required to distribute accrued benefits equally, and benefit sharing is the demand for tangible benefits to local communities living within or near protected areas. Benefit sharing that is equitable will improve mutual understanding between local communities and conservation stakeholders. Benefits devoted to the construction of schools, hospitals, water wells, and roads appear to be less tangible to locals because what they really want is direct cash flow in their pockets. For example, in western Serengeti, the construction of schools became a motivator for poaching. Because conservationists build schools, parents must pay school fees and uniforms to send their children to these schools. Many PAs in Africa followed the Yellowstone conventional and exclusionary principle of a top-down approach, which was used during the formation of the world's oldest national park in 1872. As a result, many PAs have prioritized biodiversity and ecosystem services over social and cultural issues that are important to local communities. Poaching is an immediate option for covering these costs. In this scenario, helping societies is not always perceived as such, but rather as a threat to biodiversity [6–9].

Tanzania is endowed with an abundance of natural resources, and the country has set aside more than 30% of its land for conservation. This 30% is where 22 national parks are located; 16 game reserves; 38 game-controlled areas; 13 wildlife management areas; and 28 forest reserves. About 20% of the Tanzanian revenue comes from tourism. Therefore, fostering conservation initiatives is of paramount importance for the future conservation activities of the country. The government of Tanzania has made significant contributions to achieving the conservation vision that was laid down by Mwalimu Julius K. Nyerere [10], who was the first Tanzanian president and the father of the nation after independence in 1961 and until his retirement in 1985.

The Mwalimu conservation vision was manifested during the Arusha Manifesto in 1961, and it states:

The survival of our wildlife is a matter of grave concern to all of us in Africa. These wild creatures amid the wild places they inhabit are not only important as a source of wonder and inspiration, but are an integral part of our natural resources and our future livelihood and wellbeing. In accepting the trusteeship of our wildlife we solemnly declare that we will do everything in our power to make sure that our children's grand-children will be able to enjoy this rich and precious inheritance. The conservation of wildlife and wild places calls for specialist knowledge, trained manpower and money, and we look to other nations to co-operate with us in this important task - the success or failure of which not only affects the continent of Africa but the rest of the world as well.

The Mwalimu conservation vision was a first-of-its-kind, symbolic plan that was put into action in a strong way. Its goal was to protect and care for Tanzania's beautiful wildlife as a top national heritage [10]. In the pursuit of conservation initiatives to make this wildlife conservation paradigm come true, during 2020, the country made a substantive recategorization of protected areas, making a total of 22 national parks from only one national park, the Serengeti, since 1959. The recategorization process upgraded the status of six game reserves to a more protected category, that of a national park. The newly established national parks are: Nyerere, Burigi-Chato, Kigosi, Ibanda-Kyerwa, Rumanyika-Karagwe, and Ugalla River [11].

2. Ngorongoro landscape

Ngorongoro Conservation Area (NCA) is classified as category V by the IUCN. Category V is defined as an area with distinct ecological, biological, cultural, and scenic value as a result of long-term human–nature interaction. As a result, the area and its associated features are protected and preserved. The NCA is an example of a diverse landscape in which humans (the Maasai tribe) coexist with wildlife. The Maasai of the Ngorongoro Crater were evicted from Serengeti, the Maasai name for an endless plain. The Maasai were evicted from the Serengeti plains by colonial rule in 1959, and the area was gazetted as a national park [8, 12]. As a conservation model, colonial rule used "fines and fences," an approach that excluded people from conservation efforts, resulting in hostility between the two actors [2]. Following that, the Maasai were relocated to the Ngorongoro landscape, and they remained there since their exodus from the Serengeti plain. They had a population of about 3000 people when they relocated to NCA. They joined a small group of local Maasai at NCA landscape. According to current projections, their population on the landscape is around 120,000 people (**Figure 1**). The increase in people is proportional to the increase in livestock, as the Maasai who live in the Ngorongoro landscape rely solely on animal husbandry as their main source of income (**Figures 1** and **2**). The 1959 Ngorongoro Conservation Ordinance strictly prohibits farming in the area (**Figure 3**) [13, 14]. As the human population grows, it becomes a burden on our mother earth. Not only that, but given the constraints imposed on the Maasai, who are limited to only one source of economic activity (animal husbandry). As a result, the government was forced to take some steps in order to secure arable land in Handeni, Tanga region (**Figure 4**). Furthermore, because they lived in the Ngorongoro landscape, the 1959 Ngorongoro Conservation Ordinance prohibited them from constructing modern homes. Given the restrictions put in place to protect the landscape's beauty, these restrictions hurt the Maasai people.

In the NCA, human and livestock population growth has a significant impact on wildlife and their habitats (**Figures 1** and **5**). Human population growth and associated activities result in significant changes in land use in any landscape (**Figure 1**). As a result, wildlife's available vegetation and habitat ranges are shrinking [15–17]. The lion population, for example, has declined significantly in the NCA (**Figure 5**). Because of increased human activity, the Serengeti National Park is now a source of predators, while the protected areas surrounding it are sinks [15]. The human population is rapidly growing in the area (**Figure 1**). As the human population grows, so does the demand for land for livestock grazing (**Figure 2**), farming, and settlement. As a result, wildlife habitats are deteriorating and being displaced by other activities [18, 19]. Some writers stated in the Cattle Complex Theory, in the context of pastoralists and their livestock, that it was difficult to draw a line between a herd of

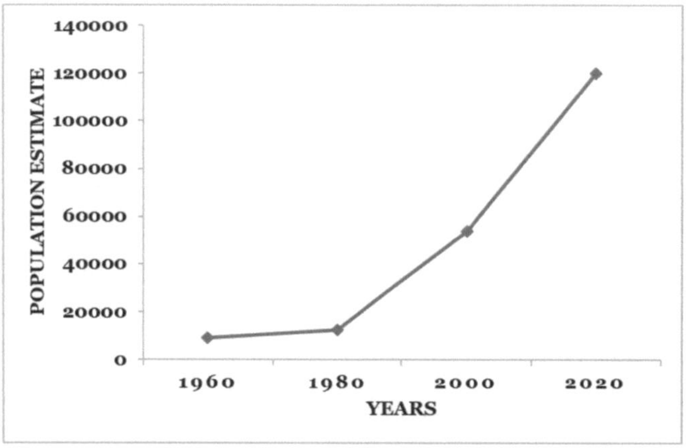

Figure 1.
Human population projections for the Ngorongoro conservation area from 1960 to 2020.

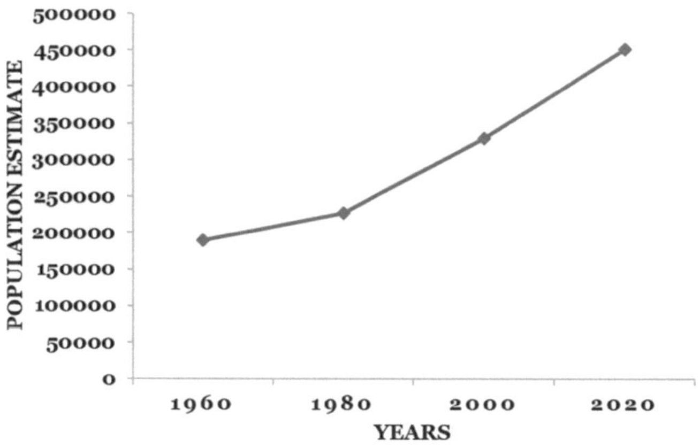

Figure 2.
Livestock population (cattle, sheep, and goats) projections for the Ngorongoro conservation area from 1960 to 2020.

cattle and the cattle herder (difficulty distinguishing master from slave). Eventually, they concluded that cattle are in charge in African pastoralism because herders try to meet their needs for food, water, space, and disease protection [20]. The Ngorongoro district currently has approximately 419,355 hectares suitable for cattle grazing. Tanzania has 12, 513 villages, but only 648 villages have been authorized to dedicate their land for grazing [20].

Human population growth in many African countries has increased the demand for land use and natural resource exploitation significantly. As the human population grows, especially in Africa, wild prey species such as wildebeest are depleted (**Figure 6**) and natural habitats are displaced; these habitats are now limited to small patches [21]. As a result, a proper management structure and policy designed to conserve natural habitats and their species is urgently required [22, 23]. When humans dominate the areas surrounding protected areas, their activities have a huge negative impact on wildlife [21]. In Algonquin Park, Canada, for example, the wolf population was severely persecuted when the animals crossed the park border [24]. Because of human population growth and the expansion of agricultural activities for global food security, biodiversity is

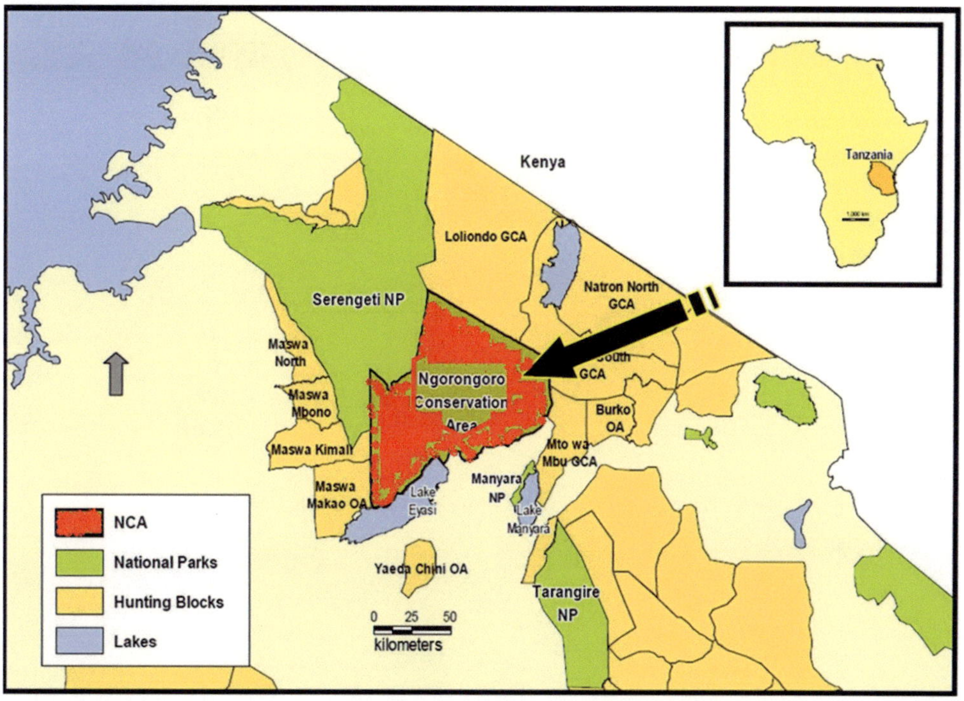

Figure 3.
The Maasai people' current home, with red highlights outside the crater.

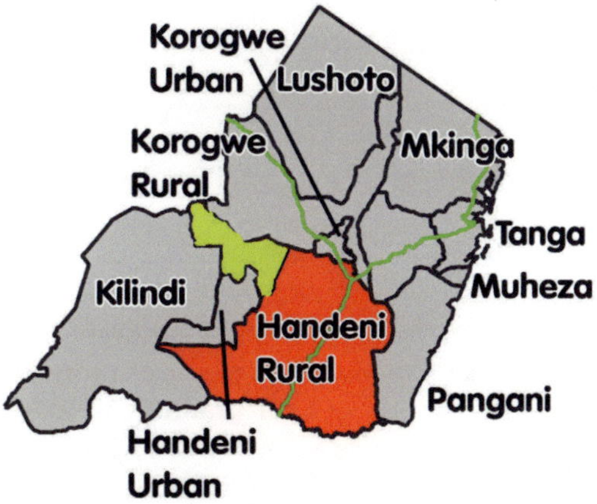

Figure 4.
Resettlement zone in light green.

currently facing a significant challenge. Anthropogenic activities have significantly altered land cover and land use, which has a negative impact on biodiversity in any ecosystem [25]. For example, in the Serengeti ecosystem, large carnivore populations are higher in the Serengeti National Park than in the Loliondo Game Controlled Area (LGCA) [26], whereas large carnivore abundance was similar in the two areas in previous decades (**Figure 5**) [18]. This shift is the result of an increase in pastoral community populations that live within the LGCA with their livestock.

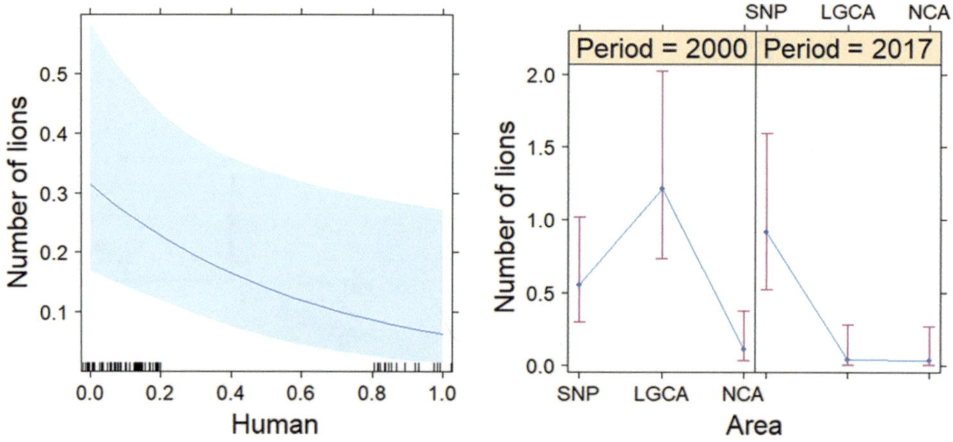

Figure 5.
Lion population status in the Ngorongoro conservation area for the two periods (2000–2017) [15].

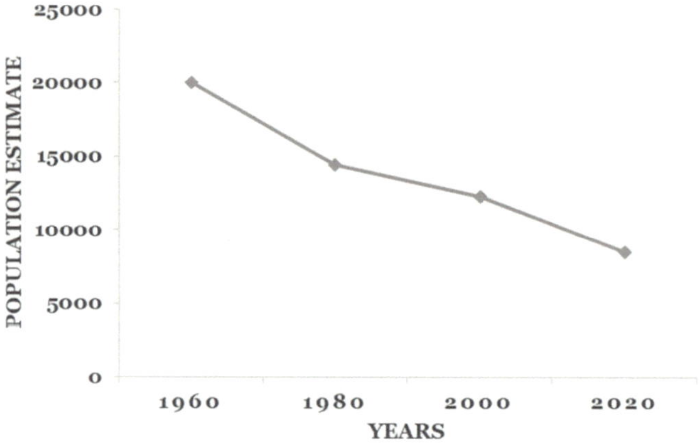

Figure 6.
Wildebeest population decline in the Ngorongoro conservation area from 1960 to 2020.

Tanzania's government, through the Ministry of Natural Resources and Tourism, is doing everything possible to keep the Mwalimu vision of protecting wild places alive for present and future generations. Today, high attention is focused on the decline and extinction of species, habitats, and entire ecosystems. The most serious attention is given to the loss of biodiversity due to species extinction. Therefore, on the good will of the government of Tanzania to make sure that its people live in a peaceful place with freedom to legally perform daily economic activities, the Maasai were voluntarily asked to register and relocate to Handeni, Tanga. In the new home, the government built modern homes and provided infrastructure for social services such as hospitals, water wells, electricity supply, and road networks, to mention a few. The exercise is still in progress as many people are still willing to relocate to the area (**Figure 7**). The public is still unaware of the number of new houses that are expected to be built, but it may be determined by how many people are willing to relocate to the area. I understand how difficult it is to relocate 120,000 people, and because the exercise is based on individual willingness to relocate, it is anticipated that not all Maasai will leave the NCA landscape (personal observation).

Figure 7.
The current housing style of the Maasai people at Ngorongoro.

3. Livestock depredation and human attack incidences in the Ngorongoro landscape

Globally, wildlife habitats are gradually shrinking as a result of anthropogenic activities and human population growth [23]. Because of an increased human population, the edges of protected areas are gradually shrinking, resulting in an increased demand for land for settlement and farming. Such demands have a negative impact on biodiversity conservation by encroaching on arable and fertile lands adjacent to protected areas [27]. As a result of habitat degradation, frequent interactions between humans and wildlife have increased [21, 23]. Because human population control in Africa is poor, there is a strong correlation between the increase in human population and the extinction of wildlife [28, 29]. Large carnivores and humans have a long history of conflict, with human attacks and livestock depredation [30, 31]. Human attacks and livestock predation have resulted in carnivore retaliation [31, 32]. This conflict must be addressed using all available means in order to gain conservation support for large carnivores from local people who interact with these species [23, 33]. For example, a study of the feeding and seasonal preferences of lions (*Panthera leo*) in Kenya's Tsavo National Park discovered that livestock depredation increased during the dry season [34]. In contrast to the dry season, when most herds of prey are found around water points, carnivores' prey is evenly distributed across the habitat during the wet season. This facilitates predation because lions can hide near water sources and prey on approaching herds [34]. However, chasing individuals during the wet season requires more effort and energy due to the abundance of shallow water points. When predators find it difficult to hunt wild prey, they turn to easier prey, such as livestock [34]. Furthermore, it was discovered that livestock depredation increased during the wet season in the Tsholotsho Communal Land and Ngamo Forest adjacent to Hwange National Park in Zimbabwe [35]. Communities usually grow crops during the wet season. During this time, their livestock is moved away from the crop fields to graze near protected areas where there are more wild carnivores [35]. When dealing with livestock depredation, coexistence measures are easier to implement in a community than human attacks. It is frequently possible to provide conservation incentives that boost local morale and increase their willingness to accept losses as a result of the incentives. It is, however, impossible to provide incentives that compensate for human attacks and/or killings. Ideally, studies should be conducted to educate people on how to avoid human attacks based on the circumstances surrounding previous attacks. The study's findings will be used to propose various strategies for harmonious coexistence between humans and carnivores as well as to support their conservation. Tanzania's conservation stakeholders' primary goal is to reduce the carnivore population adjacent to many networks of protected areas. Retaliatory killings will increase if people have negative attitudes toward large carnivores. Large carnivores require large home ranges, and because protected areas are shrinking due to anthropogenic activities, they end up living close to human-dominated areas.

The situation in the Ngorongoro landscape is threatening people's lives due to frequent incidents of human attacks and/or livestock depredation. Normal lives of people living in the landscape have been jeopardized, especially for schoolchildren, who now attend school in comparison to previous generations when many Tanzanians did not. When students arrive at school in the morning, they are vulnerable to predator attacks. With all of this chaos, only a few families are brave enough to send their children to school. With the new home in Handeni, the constraints that would have resulted in an illiterate society in the future are now a thing of the past.

4. Improving the quality of the Maasai people with modern homes and farming opportunities

A large number of Maasai communities have voluntarily registered to relocate to the newly established homes in Tanga's Handeni district [36, 37]. It is time to educate the public about the distinctions between the NCA and the LGCA) in the Loliondo and Sale divisions. According to the IUCN, NCA are classified as category V. However, the Wildlife Division is in charge of the LGCA. It is a gazetted area with defined boundaries where people are permitted to live. A permit or license is required to exploit or harvest wildlife resources. Some species are not hunted, for example, rhinos and giraffes, as well as there is wildlife control for crop or livestock protection. So, the NCA only lets people visit for photography, while the LGCA lets people visit for both photography and trophy hunting.

When discussing the Ngorongoro district, where both the NCA and the LGCA are located, it is common to become confused. As a result, they are found in the same district but play different roles in terms of wildlife conservation and tourism businesses. Tanzania's government believes that they have the right to protect their people from any danger. In this context, the government defines the danger as human attacks, livestock depredation, and a poor quality of life [31, 38] as well as the consequences of raising an illiterate generation because most children do not attend school due to fear of lion attacks [39]. Tanzania's government has taken steps to build houses in the Tanga region (Handeni district) (**Figure 4**). The Handeni district appears to be the ideal location for Maasai people and livestock in order to alleviate current pressures on the well-known NCA [40, 41]. The presence of NCA contributes to a higher percentage of tourism revenue generated by tourism activities. As a result, the Tanzanian government has determined that the only way to protect the status and scenic beauty of the NCA is to ask the Maasai people to relocate to a safe place (Handeni district) that is free of human attacks caused by large carnivores and/or livestock depredation. The artifact of constructing new modern homes in better conditions than existing traditional houses is in accordance with the country's goal of continuously improving the quality of life of its people. As a result, the majority of Tanzanians see the government's actions as beneficial to Maasai communities. Maasai people in NCA currently live in boma houses made of poles and cow dung and thatched with grass or occasionally covered with plastic [42] (**Figure 3**). The conservation policy that governs the NCA prohibits the Maasai from engaging in farming activities [43, 44]. The only advantage they have is free pasture, and the remainder of their food should come from nearby districts [43, 45]. The new home, Handeni, on the other hand, has no such restrictions. In this anticipation, it is a good opportunity for each household to have a good house made of bricks and roofed with iron sheets (**Figure 8**). Furthermore, they will have a backyard to grow food crops such as maize and vegetables to supplement their diet. What a fantastic opportunity!

Nevertheless, this opportunity may prove difficult during their initial days in their new home. For centuries, the Maasai have been pastoralists, relying on livestock husbandry for a living. Outside the NCA, the Maasai practice both livestock husbandry and subsistence farming, like in the LGCA [17]. Since the Maasai from NCA have not done farming before, they will need help and training from extension officers from the Ministry of Agriculture.

Figure 8.
The new housing style of the Maasai people at Handeni.

5. Conclusion

Given the current unprecedented increase in human population, now is the time for conservationists and policymakers to rethink and find immediate solutions to conserve and protect biodiversity. If we do not plan for the future, the natural wonders we enjoy today will be lost and our grandchildren will blame us for irresponsible actions. As the Mwalimu (the father of Tanzania) vision emphasizes, the loss of these magnificent wild places and their creatures will affect not only the African continent but also the entire world. As a result, Tanzania's decision to willingly allow the Maasai to relocate to freedom land was the millennia decision ever made in conservation initiatives. For instance, according to the Martin Oleikayo, the head of Msomera Village, said that "the resettlement plan is good for the whole country." He also said that "all citizens benefit from the money made by tourism in the conservation area" [36]. A bold decision on economic potential due to the area's growing human population will result in a medium- and eventually high-income populace. With the present climate change and population pressures, it will aggravate poverty in the region [46]. However, the Maasai resettlement may have some drawbacks. It is very likely that moving out of the area where you have lived since you were a child will cause some discomfort. The trade-offs that are being thought of here are ones in which the option of moving is better than the option of staying on their ancestral land.

Author details

Franco P. Mbise
Department of Biology, The University of Dodoma, Dodoma, Tanzania

*Address all correspondence to: francombise@gmail.com;
franco.mbise@udom.ac.tz

References

[1] Dudley N, Parrish JD, Redford KH, Stolton S. The revised IUCN protected area management categories: The debate and ways forward. Oryx. 2010;**44**(4):485-490. DOI: 10.1017/S0030605310000566

[2] Kideghesho JR, Røskaft E, Kaltenborn BP. Factors influencing conservation attitudes of local people in Western Serengeti, Tanzania. Biodiversity and Conservation. 2007;**16**(7):2213-2230. DOI: 10.1007/s10531-006-9132-8

[3] Dembe E, Bergin P. Defining the "conservation" in community conservation: Strategic planning in Tanzania National Parks. African Wildlife Foundation Discussion Papers Series, CC-DP-5, No. 5. 2013

[4] Campbell LM, Vainio-Mattila A. Participatorydevelopmentandcommunity-based conservation: Opportunities missed for lessons learned? Human Ecology. 2003;**31**(3):417-437. DOI: 10.1023/A:1025071822388

[5] Anthony B. The dual nature of parks: Attitudes of neighbouring communities towards Kruger National Park, South Africa. Environmental Conservation. 2007;**34**(3):236-245. DOI: 10.1017/S0376892907004018

[6] Figueroa E, Aronson J. New linkages for protected areas: Making them worth conserving and restoring. Journal for Nature Conservation. 2006;**14**(3-4):225-232. DOI: 10.1016/j.jnc.2006.04.007

[7] Htay T, Htoo KK, Mbise FP, Røskaft E. Factors influencing communities' attitudes and participation in protected area conservation: A case study from northern Myanmar. Society and Natural Resources. 2022;**35**(3):301-319. DOI: 10.1080/08941920.2022.2032515

[8] Newmark WD, Hough JL. Conserving wildlife in Africa: Integrated conservation and development projects and beyond: Because multiple factors hinder integrated conservation and development projects in Africa from achieving their objectives, alternative and complementary approaches for promoting wildlife conservation must be actively explored. Bioscience. 2000;**50**(7):585-592. DOI: 10.1641/0006-3568(2000)050[0585:CWIAIC]2.0.CO;2

[9] Egli L, Meyer C, Scherber C, Kreft H, Tscharntke T. Winners and losers of national and global efforts to reconcile agricultural intensification and biodiversity conservation. Global Change Biology. 2018;**24**(5):2212-2228. DOI: 10.1111/gcb.14076

[10] Burnett GW, Conover R. The efficacy of Africa's National Parks: An evaluation of Julius Nyerere's Arusha manifesto of 1961. Society and Natural Resources. 1989;**2**(1):251-260. DOI: 10.1080/08941928909380690

[11] TANAPA. Ministry of Natural Resources and Tourism: Tanzania National Parks Authority. 2022. Retrieved from: https://www.tanzaniaparks.go.tz/

[12] Kideghesho JR. 'Serengeti shall not die': Transforming an ambition into a reality. Tropical Conservation Science. 2010;**3**(3):228-247. DOI: 10.1177/194008291000300301

[13] NBS. Human Population Trend by the National Bureau of Statistics. Tanzania. Retrieved from: https://www.nbs.go.tz/: Dar es Salaam; 2022

[14] NCA. Ministry of Natural Resources and Tourism: Ngorongoro Conservation

Area Authority. Arusha, Tanzania: World Heritage Centre Report; 2010

[15] Jackson C et al. Census of Carnivores, Vultures and Other Raptors Using Audio Playbacks. Brussels, Belgium: European Commission for Research and Innovation. Ref. Ares (2019)6579269-24/10/2019. https://ec.europa.eu/research; 2019

[16] Veldhuis MP et al. Cross-boundary human impacts compromise the Serengeti-Mara ecosystem. Science. 2019;**363**(6434):1424-1428. DOI: 10.1126/science.aav0564

[17] Mbise FP, Jackson CR, Lyamuya R, Fyumagwa R, Ranke PS, Røskaft E. Do carnivore surveys match reports of carnivore presence by pastoralists? A case of the eastern Serengeti ecosystem. Global Ecology Conservation. 2020;**24**:e01324. DOI: 10.1016/j.gecco.2020.e01324

[18] Maddox TM. The Ecology of Cheetahs and Other Large Carnivores in a Pastoralist-Dominated Buffer Zone. Department Anthropology. PhD thesis. 2003. p. 373

[19] Lyamuya RD, Masenga EH, Mbise FP, Fyumagwa RD, Mwita MN, Røskaft E. Attitudes of Maasai pastoralists towards the conservation of large carnivores in the Loliondo game controlled area of northern Tanzania. International Journal of Biodiversity Conservation. 2014;**6**(11):797-805. DOI: 10.5897/IJBC2014.0769

[20] Onesmo OP. An Asssesment of the Tanzania Livestock Policy of 2006 and its Impact on the Livelihoods of Pastoral Communities in Tanzania. Master thesis. Tanzania: University of Dar es Salaam; 2016

[21] Yirga G et al. Local spotted hyena abundance and community tolerance of depredation in human-dominated landscapes in northern Ethiopia. Mammalian Biology. 2014;**79**(5):325-330. DOI: 10.1016%2Fj.mambio.2014.05.002

[22] Hazzah L, Bath A, Dolrenry S, Dickman A, Frank L. From attitudes to actions: Predictors of lion killing by Maasai warriors. PLoS One. 2017;**12**(1):e0170796. DOI: 10.1371/journal.pone.0170796

[23] Ronnenberg K, Habbe B, Gräber R, Strauß E, Siebert U. Coexistence of wolves and humans in a densely populated region (Lower Saxony, Germany). Basic and Applied Ecology. 2017;**25**:1-14. DOI: 10.1016/j.baae.2017.08.006

[24] Woodroffe R, Frank LG. Lethal control of African lions (Panthera leo): Local and regional population impacts. Animal Conservation forum. 2005;**8**:91-98. DOI: 10.1017/S1367943004001829

[25] Wessels KJ, Reyers B, Van Jaarsveld AS. Incorporating land cover information into regional biodiversity assessments in South Africa. Animal Conservation forum. 2000;**3**:67-79. DOI: 10.1111/j.1469-1795.2000.tb00088.x

[26] Craft ME, Hampson K, Ogutu JO, Durant SM. Carnivore communities in the greater Serengeti ecosystem. Serengeti IV Sustainability Biodiversity Coupled Human-Nature System. 2015;**11**:419-447

[27] Nyhus PJ, Tilson R. Panthera tigris vs Homo sapiens: Conflict, coexistence, or extinction. In: Tigers of the World. Amsterdam, The Netherlands: Elsevier; 2010. pp. 125-141

[28] Linnell JDC, Swenson JE, Andersen R. Predators and people: Conservation of large carnivores is possible at high human densities if mangement policy is favourable. Animal Conservation. 2001;**4**:345-349

[29] Songorwa AN. Human population increase and wildlife conservation in Tanzania: Are the wildlife managers addressing the problem or treating symptoms? African Journal of Environmental Assessment and Management. 2004;**9**:49-77

[30] Packer C, Ikanda D, Kissui B, Kushnir H. Lion attacks on humans in Tanzania. Nature. 2005;**436**(7053):927-928. DOI: 10.1038/436791a

[31] Ikanda D, Packer C. Ritual vs. retaliatory killing of African lions in the Ngorongoro conservation area, Tanzania. Endangered Species Research. 2008;**6**(1):67-74. DOI: 10.3354/esr00120

[32] Kissui BM. Livestock predation by lions, leopards, spotted hyenas, and their vulnerability to retaliatory killing in the Maasai steppe, Tanzania. Animal Conservation. 2008;**11**(5):422-432. DOI: 10.1111/j.1469-1795.2008.00199.x

[33] Gurung B, Smith JLD, McDougal C, Karki JB, Barlow A. Factors associated with human-killing tigers in Chitwan National Park, Nepal. Biological Conservation. 2008;**141**(12):3069-3078. DOI: 10.1016/j.biocon.2008.09.013

[34] TLP. Tsavo Lion Project: Long-term study of the ecology and environmental preferences of the lions in the greater Tsavo region. 2018. Retrieved from: https://lionconservationfund.org/tsavo.html

[35] Kuiper TR et al. Seasonal herding practices influence predation on domestic stock by African lions along a protected area boundary. Biological Conservation. 2015;**191**:546-554. DOI: 10.1016/j.biocon.2015.08.012

[36] NCA. Ministry of Natural Resources and Tourism: Ngorongoro Conservation Area Authority. 2022. Retrieved from: https://www.ncaa.go.tz/

[37] MNRT. United Republic of Tanzania: Ministry of Natural Resources and Tourism. 2022. Retrieved from: https://www.maliasili.go.tz/

[38] Mbise FP. Attacks on humans and retaliatory killing of wild carnivores in the eastern Serengeti ecosystem, Tanzania. Journal of Ecology Nature Environment. 2021;**13**(4):110-116. DOI: 10.5897/JENE2021.0908

[39] Lyamuya RD et al. Can enhanced awareness change local school children's knowledge of carnivores in northern Tanzania? Human Dimensions of Wildlife. 2016;**21**(5):403-413. DOI: 10.1080/10871209.2016.1180566

[40] Galvin KA, Thornton PK, Boone RB, Knapp LM. Ngorongoro conservation area, Tanzania: Fragmentation of a unique region of the greater Serengeti ecosystem. In: Fragmentation in Semi-Arid and Arid Landscapes. Switzerland: Springer Nature; 2008. pp. 255-279

[41] Lawuo ZA, Mbasa B, Mnyawi S. Persistence of land conflicts between maasai community and ngorongoro conservation area authority (NCAA) in ngorongoro conservation area (NCA). International Journal of Innovation Science Research. 2014;**5**(2):154-161

[42] Mbise FP, SkjÃ GR, Lyamuya RD, Fyumagwa RD, Jackson C, Holmern T. Livestock depredation by wild carnivores in the eastern Serengeti ecosystem, Tanzania. International Journal of Biodiversity Conservation. 2018;**10**(3):122-130. DOI: 10.5897/IJBC2017.1165

[43] Kivelia J. Cultivation Trends in the Buffer Zones of East African Rangeland Protected Areas: The Case of Ngorongoro and Loliondo in Tanzania. University of London, University College London (United Kingdom); 2007

[44] Millanga P. Conflicting Land Use and Management Strategies in the Ngorongoro Conservation Area (NCA) of Northern Tanzania. Carleton University; 2009

[45] Daniel G. Assessment on Impact of Human Population Pressure on

the Sustainabilty of Ngorongoro Conservation Area. The University of Dodoma; 2015

[46] Slootweg S. Climate Change and Population Growth in Pastoral Communities of Ngorongoro District, Tanzania. Handbook of Climate Change Resilience. Switzerland: Springer Nature; 2018. pp. 1627-1645

Section 2

Conservation Biology

Chapter 5

Natural and Anthropogenic Impacts on the Macrophytes of Soft-Water Lakes of Estonia

Helle Mäemets

Abstract

Oligotrophic and semidystrophic lakes (water alkalinity <80 mg HCO_3^- L^{-1}; dichromate oxygen consumption <40 mg O L^{-1}) are the main habitats for rare macrophytes, especially isoetids. They are characteristic for the lakes with HCO_3^- $\leq$ 30 mg L^{-1}; the higher alkalinity and related higher trophy level support elodeids. Anthropogenic impact on these lakes in Estonia started with flax retting and water lowering and continued with sauna building, agricultural nutrients, and holiday activities. The present overview is based on the data of the last 50 years. Anthropogenic acidification of Estonian lakes is not known, but natural dystrophication due to the inflow of humic compounds is probable. Alkalization and eutrophication are closely related, amplifying each other, and water level modifies these processes. Eutrophication increases the occupation of shallow zone by emergent belts, suppression of isoetids by elodeids, and overshadowing by phytoplankton blooms in a deeper zone or host plants by macroalgae. Fast-growing eutraphents accelerate the accumulation of organic sediments, unfavorable for isoetids. Among floating-leaved plants hybridization between rare and common species as well as introgression takes place. It is almost impossible to reverse back ecosystems that formed and balanced over thousands of years and became unbalanced during a much shorter period. Easier would be to keep functioning ecosystems.

Keywords: oligotrophy, dystrophication, eutrophication, hydrophytes, competition

1. Introduction

The concept of oligotrophic state of the lakes has been applied relatively widely, mostly not concerning natural types of lakes. In literature, it has been used also for the lakes with mean or high content of the bicarbonate ion (HCO_3^-) in the water, if the content of main nutrients (C, N, P) is low. Light soft-water lakes are 'real' oligotrophic (OL) lakes, because in some cases, their clear water is comparable with distilled water. The next step of their development may be semidystrophic (SD) lakes, where moderate content of humic compounds from the forest or peatland stains water yellow or light brown [1]. Most of these two types, hereafter, named soft-water lakes (SWL—excluding acidotrophic and dystrophic lakes), are located in the boreal region of the northern hemisphere, but they occur also in warmer climates, for example, in Atlantic

IntechOpen

dyne areas of SW France, as well as in mountains. For the composition of macrophytes and other primary producers, the content of bicarbonate ions in the water plays a significant role, allowing the growth of HCO_3^- demanding macrophytes in SWL only close to inflows of mostly hard (with some exceptions) groundwater in the littoral. Long-term research in North Wisconsin revealed that, in seepage lakes, groundwater recharge formed 0–25% [2]. When the nutrient loading is accompanied by bank erosion or a decline in water level (an increase of the proportion of groundwater), HCO_3^- supply increases too.

Most characteristics of SWL are submerged mosses and isoetids that are small, slow-growing and adapted to the nutrient-poor environment. However, adaptation to the nutrient-poor habitat does not mean that these plants perish immediately after uncommon nutrient loading. Mostly they are suppressed during the following years, by fast-growing 'common' species, being able to colonize the habitat in changed conditions and producing larger amounts of organic remains leading to oxygen depletion. The other negative impact of higher trophy for submerged plants is mass-propagation of epiphytic and/or planktonic algae and a decline in light availability. In some taxa, for example, *Nuphar* and *Sparganium*, eutrophication supports introgression, while common species hybridizes with the specimens of rare species and again with the hybrids during the generations.

Estonian soft-water lakes are formed during the Late Glacial and Holocene in sandy areas on both alkaline (limestone) and acidic (sandstone) bedrock (**Figure 1**). Sandy areas were formed by glacier meltwaters or on coastal terraces of Baltic Sea [3], and during the Holocene sand in the slopes of the lake basins was washed through,

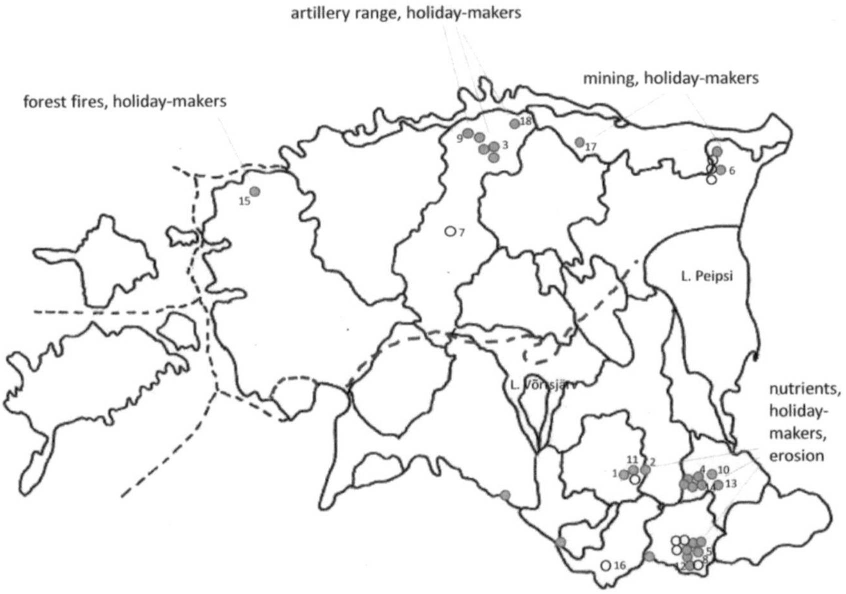

Figure 1.
Main anthropogenic impacts on the OL and SD lakes in different regions of Estonia and the occurrence of rare macrophytes: empty circles—found in 1971–2011; blue circles—persisted also in 2012–2022; brown dashed line—approximate border between Ordovician & Silurian limestones and Devonian sandstone. Black lines—geographical (landscape) districts. The numbers of lakes in ***Table 1****: 1, Hüüdre; 2, Ihamaru Palojärv; 3, Jussi Pikkjärv; 4, Karsna; 5, Kirikumäe; 6, Kurtna Valgejärv; 7, Matsimäe Pühajärv; 8, Mikeli; 9, Mähuste; 10, Nohipalu Valgjärv; 11, Piigandi; 12, Pullijärv; 13, Solda; 14, Tsolgo Pikkjärv; 15, Tänavjärv; 16, Ubajärv; 17, Uljaste; 18, Viitna Pikkjärv.*

reducing their nutrient content. General delimitation of SWL according to alkalinity of the epilimnetic layer in summer is <80 mg HCO_3^- L^{-1}. These lakes are divided into two groups: (A) alkalinity <30 mg; (B) 30–80 mg HCO_3^- L^{-1} [4]. The content of dissolved organic matter/humic compounds, expressed as chemical oxygen demand: dichromate oxygen consumption (COD) is mostly <35 mg O L^{-1}. The lakes with remarkably higher COD belong to dystrophic (> 40) and acidotrophic (> 80 mg) groups, [1, 4], not addressed below. Due to negligible buffer capacity, OL lakes are most sensitive to eutrophication, and nowadays in Estonia oligotrophic lakes of reference state are lost. On the basis of the data from the twentieth century, the book concerned on Estonian SWL in 1991 covered 121 lakes [4], but among them about 50 were already eutrophic. At the end of the twentieth century, an estimated number for the lakes close to OL was 20, and for SD lakes, 40 [1]. These two lake types are the main habitats for several rare hydrophytes and have been subject to the both, natural and mediated/immediated anthropogenic impacts. The aims of the present overview is to discuss already specified threats to SWL and to point out particular features of Estonian SWL. Hopefully, it helps to understand which changes may be irreversible and what is in our chancery to improve the situation.

2. Materials

First single descriptions of the vegetation of small, < 100 km^2 lakes, including soft-watered, date back to the beginning of the twentieth century, and some lake groups in East Estonia were studied already between two World Wars. The main part of investigations were carried out since the 1950s, and today most of the SWL have been studied repeatedly but not regularly. Macrophyte data are assembled into a special database and kept at the Chair of Hydrobiology and Fishery of Estonian University of Life Sciences. There, the data from 1960s to 1990s originate from the studies by Aime Mäemets (1930–1996); subsequently, SWL were studied by the author, L. Freiberg, Dr. K. Palmik-Das, G. Ratasepp, Dr. K. Karus, Dr. T. Feldmann and M. Lehtpuu. A more intense investigation of the macrophytes of SWL started in 2011 when the author studied 52 lakes for the compiling of protection plans for rare macrophytes. Later on, inventories of protected plants were carried out in 2012 (18 SWL), in 2018–2019 (26 SWL) and in 2021–2022 (20 SWL) by the author, Dr. K. Palmik-Das and MSc student E. Kara.

Passing the littoral of the whole lake by boat, composition and relative abundances of the macrophyte species (separately for all belts: emergent, floating-leaved, etc.) were estimated for a lake in a modified 1 ... 5 scales of Braun-Blanquet. Using the plant hook with marked rope (or stick), depth limits of macrophytes were measured on transects. Also, vegetation schemes were drawn in the field, using the signs of taxa. This method has been used since the 1950s and serves as the basis for the estimation of ecological quality classes for SWL [5]. At the inventories of protected macrophytes also an approximate number of their individuals was estimated, as well as GPS coordinates and distribution polygons were fixed. Additionally, vitality of plants, entailed taxa and other circumstances were studied.

The area of the 60 OL and SD lakes varies between 1 and 137 ha. However, the majority—38 lakes—belong to the group of <10 ha water bodies, overwhelmingly seepage lakes. Macrophyte data with different frequencies are available for all lakes. There are some deeper and some very small A-lakes without any submerged vegetation during the all recorded time, very probably due to flax retting, taking place (mainly) until the first half of the twentieth century. Preferred were SWL, and at least 15 SWL have 'flax' in their name. We do not know about their vegetation before flax retting.

As shown in **Figure 1**, during the last 50 years, rare macrophytes were found in 38 lakes [6], but in the last decade, they have not more occurred in nine lakes. In 29 lakes with the newest findings, vital populations of at least one species occurred in 16 lakes. The map in **Figure 1** does not include the lakes where during the studied period only very few vegetative (undetermined) narrow-leaved *Sparganium* plants were found.

Hydrochemical data of 18 SWL inhabited by rare plants are presented in **Table 1** and the list of characteristic vascular plants of SWL in **Table 2**. Among selected lakes annual hydrochemical monitoring of the lakes of Nohipalu Valgjärv, Uljaste and

Lake	Year	SD m	pH	HCO_3^- mg L^{-1}	COD mg O L^{-1}	P_{tot} mg P L^{-1}	N_{tot} mg N L^{-1}
Hüüdre (Sg)	1972	1.8	6.8	12.2	46	—	—
Hüüdre (**Sg***)	2021	0.5	9.0	13.1	65	0.057	1.600
Ihamaru Palojärv (**Il Sa**)	1964	3.3	6.0	0.01	33	—	—
Ihamaru Palojärv (**Il Sa**)	1981	3.5	6.4	4.27	41	—	—
Ihamaru Palojärv (Il)	2005	2.6	8.0	9.15	26	0.015	0.683
Jussi Pikkjärv (Sa N)	1990	2.0	7.0	3.0	29	0.030	0.300
Jussi Pikkjärv (Sa N)	2009	2.7	6.9	≤6.1	28	0.025	0.429
Jussi Pikkjärv (Sa N)	2021	3.3	6.5	5.30	23	0.013	0.530
Karsna (**Il Sg E**)	1981	4.0	7.1	30.5	54	—	—
Karsna (**Sg** Il Ma)	2012	1.7	6.2	12.2	44	0.028	1.000
Karsna (**Ma Sg*** Il)	2021	1.9	7.8	15.2	40	0.026	1.100
Kirikumäe (**L Il**)	1990	1.0	6.4	3.1	48	0.094	1.830
Kirikumäe (**L Il**)	2010	1.4	7.5	3.1	32	0.025	0.520
Kurtna Valgejärv (**L** Il)	1981	4.3	7.0	18.3	26	—	—
Kurtna Valgejärv (**L** Il)	2001	2.1	6.7	6.0	—	0.008	0.390
Kurtna Valgejärv (**L** Il, Sa)	2010	1.9	7.6	<6.1	42	0.015	0.420
Matsimäe Pühajärv (Sa)	1964	1.3	≤5	0	43	—	—
Matsimäe Pühajärv	2021	—	4.8	3.04	40	0.026	1.100
Mikeli (Sa)	1970	2.0	5	—	—	—	—
Mikeli	2021	—	4.9	≤1	80	0.047	0.760
Mähuste (**L Il N** Ie)	1969	5.3	6.7	0.6	10	—	—
Mähuste (**L Il N** S*)	1981	3.5	6.6	18	32	—	—
Mähuste (**L Il** N Sg)	2021	3.2	6.8	5.24	28	0.013	0.670
Nohipalu Valgjärv (**Il Sa** L)	1973	4.5	6.0	1.2	14	—	—
Nohipalu Valgjärv (**Il S*** L)	1991	3.8	6.9	6.1	12	0.016	—
Nohipalu Valgjärv (**Il** S*)	2012	4.5	6.3	3.05	23	0.013	0.690
Piigandi (Il S*)	1973	4.5	7.4	24.4	25	—	—
Piigandi (**Il** Sg N)	1991	3.8	7.2	42.7	12	0.011	0.530

Lake	Year	SD m	pH	HCO_3^- mg L^{-1}	COD mg O L^{-1}	P_{tot} mg P L^{-1}	N_{tot} mg N L^{-1}
Piigandi (**Il**)	2012	3.1	8.2	48.8	—	0.015	0.500
Piigandi (**Il** S*)	2021	3.0	8.5	50.0	29	0.019	0.530
Pullijärv (**Ma** L, S*)	1973	2.9	8.0	24.0	28	—	—
Pullijärv (**Ma L** Sg Il)	1990	2.6	7.8	34.0	18	0.030	1.010
Pullijärv (**Ma L** E)	2010	2.8	8.6	42.7	30	0.017	0.690
Solda (**Sg** S*)	1974	1.1	8.4	38.4	40	—	—
Solda (Sa S*)	1981	0.8	10	67	50	—	—
Solda (Sg S*)	2021	0.7	7.4	29.9	120	0.052	1.700
Tsolgo Pikkjärv (E S*)	1961	1.7	7.0	13.4	36	—	—
Tsolgo Pikkjärv (**Ma** Sg)	2021	2.5	8.1	29.6	<15	0.025	0.880
Tänavjärv (**L**)	1976	—	7.1	12	32	—	—
Tänavjärv (**L** Il Sg S*)	1995	1.5	7.3	6.1	24	0.010	0.400
Tänavjärv (**L** S* Il)	2012	1.4	8.4	12.2	22	0.017	0.610
Tänavjärv (**L** S* Il)	2021	1.4	7.9	12.2	27	0.019	0.780
Ubajärv(**Sg** S* Ie)	1974	1.1	7.6	16.48	48	—	—
Ubajärv (**Sg**)	1978	—	6.2	12	56	—	—
Ubajärv (**Sg** S* Ie)	1983	0.6	6.5	6.1	59	—	—
Ubajärv	2021	0.4	6.7	9.1	50	0.073	1.100
Uljaste (**L** Il)	1981	1.5	6.8	15.25	28	—	—
Uljaste (**L Il**)	2000	1.8	7.4	12	34	0.030	0.560
Uljaste (**L** Il)	2012	2.4	7.6	9.15	35	0.022	0.360
Viitna Pikkjärv (**L Il Sa**)	1972	3.3	6.5	5.49	10	—	—
Viitna Pikkjärv (**L Il Sa**)	1992	4.1	6.0	3.0	13	0.016	0.644
Viitna Pikkjärv (**L Il** Sa)	2012	3.6	8.5	12.2	26	0.016	0.610

Abbreviations in brackets: L, Lobelia*; E,* Elatine*; Ie, Il,* Isoëtes*; Sa, Sg,* Sparganium*; Ma, Msi,* Myriophyllum*; N,* Nuphar. ***Bold**, taxa with abundance 3–5; S*, hybrid or vegetative. Latin species names are shown in **Table 2**.*

Table 1.
Hydrochemical data of surface layer in midsummer in SWL inhabited by populations of protected vascular plants.

Species name	Category/Red List status	Comments
Isoëtes echinospora Durieu, spiny-spored quillwort	I/RE	Extinct
Isoëtes lacustris L., lake quillwort	II/EN	In the last decade in 12 SWL
Lobelia dortmanna L., water lobelia	II/EN	Vital in 7 SWL, most damaged by holiday-makers
Sparganium angustifolium Michx., narrowleaf bur-reed	II/EN	9 (small) pure, 10 mixed/hybrid populations

Species name	Category/Red List status	Comments
S. gramineum Georgi	II/CR	1 perished pure, 14 mixed/hybrid populations
Myriophyllum alterniflorum DC, alternate water-milfoil	II/EN	In 7 SWL, including 3 very large populations
Elatine hydropiper L., eight-stamened waterwort	II/EN	In SWL rarely, main population in large L. Peipsi
Juncus bulbosus L., bulbous rush	NT	In one SWL
Potamogeton alpinus Balb., reddish pondweed	NT	In eutrophied SWL and in other lake types and rivers
Potamogeton obtusifolius Mert. & W.D.J. Koch, blunt-leaved pondweed	NT	In eutrophied SWL and other lake types
Potamogeton gramineus L., various-leaved pondweed	—	In B- and other lakes where shallow free littoral exists
Potamogeton rutilus Wolfg., Shetland pondweed	EN	In B-lakes; occurs also in other lakes with shallow free littoral

RE, regionally extinct; EN, endangered; NT, near threatened; CR, critically endangered.

Table 2.
Submerged and amphibious macrophytes of OL and SD lakes of Estonia and their actual protection status.*

Viitna Pikkjärv has been carried out since 1994, but in **Table 1** are presented 3 years from different periods, for comparison on the larger scale. A more profound comparison of their hydrochemical data with macrophyte composition deserves special analysis. Older data on hydrochemistry, until 1996, derived from a special database of the Centre for Limnology at Estonian University of Life Sciences. Latter data are compiled from the open monitoring reports by my colleagues. Data on the phosphorus and nitrogen for the studied lakes are available since the 1990s when contemporary, more reliable analyze methods were applied. For the lakes of Matsimäe Pühajärv (**Figure 2**) and Mikeli hydrochemical data are extremely poor, but they were included due to extreme conditions for observable taxa. In 2021, water samples from 14 lakes, which had the scarcest data, were chemically analyzed (project 18,745 of Estonian Environment Investment Centre). Quagmires of SWL of South-East Estonia were studied in 2013 [7] and the latest study of water mosses took place in 2014 [8]. In 2021 MSc student E. Kara and the author carried out a botanical study in 18 lakes with *Sparganium angustifolium* Michx. and/or *S. gramineum* Georgi, and (in cooperation with Tallinn University) a study of sediments in four lakes. E. Kara carried out germination experiment with *Sparganium* seeds from sediments [9].

Littorella uniflora (L.) Asch. (I category) is found only in brackish pools on the western coast of Saaremaa Island; *Subularia aquatica* L. is very probably extinct.

3. Vegetation of soft-water lakes of Estonia

The emergent vegetation belt of SWL is usually narrow or absent. If the reed belt (*Phragmites australis* (Cav.) Trin. ex Steud.) exists, it is fragmentary and sparse in the lakes of good status. The main sedges are tufted sedge *Carex elata* Bell. ex All., bottle

Figure 2.
Pine bog dominates on the banks of L. Matsimäe Pühajärv (author's photo).

sedge *C. rostrata* Stokes and slender sedge *C. lasiocarpa* Ehrh. In SD lakes peaty banks or quacking bogs in the water border yield a considerable chemical influence, increasing the content of humic compounds and decreasing the pH of water. These stretches are without isoetids, obviously not able to grow on organic-rich, oxygen-demanding loose sediments [see also 10, 11]. Isoetids grow only at sandy banks and depend on the proportion of peaty and sandy stretches (**Figure 3**). Among nymphaeids, the most frequent is broad-leaved pondweed *Potamogeton natans* L., but also amphibious bistort *Persicaria amphibia* (L.) Gray, water lilies *Nymphaea* spp. (mostly hybrids), *Nuphar lutea* (L.) Sm. and *Nuphar pumila* (Timm) DC. are not rare. Among *Sparganium* (bur-reed) occurrence of species depends on the water alkalinity, as shown in **Figure 4**, made based on the data from the 1970s to 1980s for 24 SWL [12]. Also, the composition of submerged vascular species reflects water alkalinity: SWL with typical isoetid vegetation [excluding *Littorella*—see also 13] and without elodeids containing ≤30 mg HCO_3^- L^{-1}, that is, they are A-lakes. To draw a limit for COD is more complicated. In most of the isoetid-lakes, summer COD value remains <35 mg O L^{-1}, but in some SD lakes and some years, it has been higher (**Table 1**). We can suppose that at long-lasting enrichment of A-lakes with humic substances (at COD values ≥40 mg O L^{-1}), the isoetids decline, especially *Isoëtes*, inhabiting also deeper zone >1 m. However, data in **Table 1** do not support that supposition in all cases. Very probably, high COD data of some lakes (e.g. Karsna) reflect organic pollution from settlement too. At remarkable dystrophication, few plants of *S. angustifolium* may persist in some lakes during the decades, on a single sandy place. They are very vulnerable, also because there are commonly swimming places.

Figure 3.
Lake quillwort and narrowleaf bur-reed grow near the higher, sandy banks, avoiding pine-bog water border (L. Jussi Linajärv, author's photo).

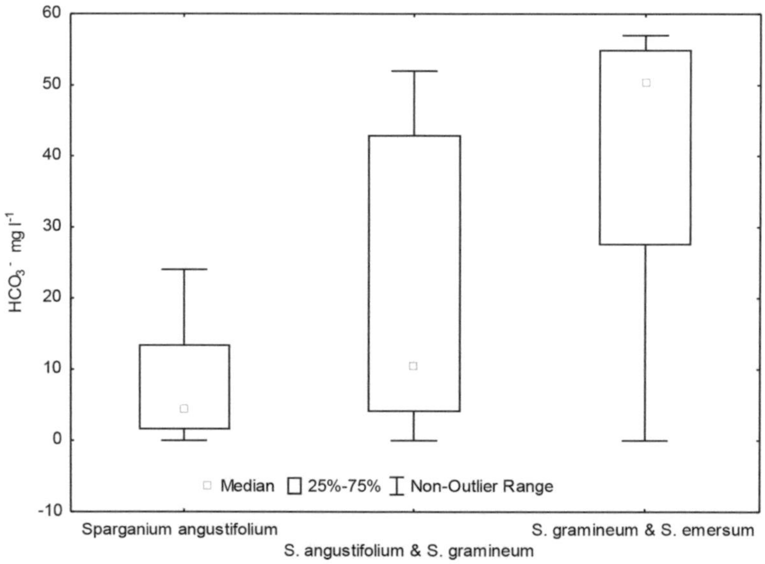

Figure 4.
Water alkalinity and Sparganium *species in 24 lakes (compiled by Dr. R. Laugaste).*

The species of submerged and amphibious vascular plants presented in **Table 2** are growing mainly in lakes in good ecological status. In the case of eutrophication, the submerged vegetation in A-lakes may disappear [6], but B-lakes may be settled by

Indicator	Unit	High	Good	Moderate	Poor	Bad
pH		5.5–7	>7–7.5	>7.5–8	>8–8.5	>8.5
Ptot	mg/l	<0.01	0.01–0.02	>0.02–0.04	>0.04–0.06	>0.06
Ntot	mg/l	<0.2	0.2–0.5	>0.5–0.8	>0.8–1.1	>1.1
SD	m	>5	3–5	2–<3	1–<2	<1

Table 3.
The limits of the values for the estimation of the ecological status of OL and SD lakes according to abiotic indicators [15].

eutraphents: fan-leaved crowfoot *Ranunculus circinatus* Sibth., Canadian pondweed *Elodea canadensis* Michx., Siberian water-milfoil *Myriophyllum sibiricum* Komar., flat-stalked pondweed *Potamogeton friesii* Rupr. etc., inhabiting zone of 1–3 m depth. These secondary taxa are not presented in **Table 2**. Emergent and floating-leaved belts increase too. Lake Piigandi (see **Table 1**) has changed from A-lake to a B-lake due to erosion of steep slopes and conjoint eutrophication, inhabiting now *Ranunculus* and *M. sibiricum*. Some elodeids, such as *Potamogeton gramineus* or *P. rutilus*, need shallow free littoral, and in B-lakes of good status, this habitat is relatively common. In the last 35 years, at least 23 species of submerged mosses have been found in SWL [8, 14], including calciphilous *Drepanocladus sendtneri* (H. Müll.) Warnst. Nowadays, submerged mosses inhabit depth zone 1–11 m [8], and their occurrence is not so clearly related to water alkalinity as to scanty availability of the main nutrients: C (mainly CO_2), N and P. Low nutrient content of the habitat is appropriate, because the metabolism of water mosses may be extremely sparing: the author has been 'cultivating' *Rhynchostegium riparioides* (Hedw.) Card. for 8 years, in a 1 l glass vessel, adding tap or rainwater and rarely removing decaying plant remains. Among charophytes *Chara virgata* Kützing and *Nitella flexilis* (L.) are not rare in B-lakes.

The main abiotic indicators of the water quality and their values for the estimation of the ecological status of OL and SD lakes, established by Estonian Ministry of Environment according to EU Framework Directive [15] are presented in **Table 3**.

4. Stressors and their synergies

As follows, the main natural and anthropogenic stressors and their interactions in SWL will be observed. Most of them are specified by Murphy [10] and Smolders et al. [16], and some are specific to Estonia.

4.1 Acidification, dystrophication and mixotrophication

Acidification in Estonian SWL, if proceeding at all, is natural, not anthropogenic, as in air-polluted Scandinavian lakes on the acidic rock. It may be caused by peaty or/and pine-forested banks and catchment. Substantially it could be dystrophication—natural development direction from OL lakes to SD lakes and further to dystrophic state [1]. Studying the quagmires of SWL in South-East Estonia [7], we found that pH in the water of quagmire societies at seven SD lakes varied largely: 3.34–7.32. In some cases, the pH in the water of moss beds differed remarkably near the same lake, for example, 3.39–6.52. The average pH was lowest in the societies of *Sphagnum fallax*

Figure 5.
Lake Ubajärv and its surroundings (clip from the map of Estonian land board). Extensive quagmire at the lake ends is probably enhanced by melioration drains (blue lines).

(3.6), *Sphagnum magellanicum* agg. & *S. fuscum* (3.7), *S. centrale* (3.8) and *S. flexuosum* (3.9) [7]. Widening of *Sphagnum* quagmires has impoverished submerged vegetation near the water border, in some SWLs, it had persisted communities of Bryales in the deeper zone. Regardless of the acidic quagmire, water pH in the surface layer of all seven lakes was ≥ 6.

For most of the lakes presented in **Table 1**, the data of COD do not support dystrophication hypothesis. Only in L. Ubajärv, surrounded by extensive forest and mire areas (**Figure 5**), a gradual increase in COD and decrease in water transparency and HCO_3^- has taken place from 1974 to 2021. *Isoëtes echinospora*, earlier occurring sporadically in the northern part, has disappeared during the last 20 years. However, this dystrophication may be enhanced by humans, by earlier melioration, in **Figure 5**, are visible the drains. High value of COD may be caused also by organic pollution from the settlement, and probably also by simultaneous input of organic matter from humic substances and settlement, resulting in enormous values, as in L. Solda (**Table 1**). If the anthropogenic inflow of nutrients overcomes associative capacity of humic substances of the lake to retain nutrients, the lake become mixotrophic: increases the amount of the nutrients in the water, phytoplankton abundance and decomposition of humic substances [1]. These processes are closely related and intensify each other.

4.2 Alkalization, eutrophication, lowering of water level

Water pH of the eutrophied SWL is a highly capricious indicator and depends on the intensity of photosynthesis, therefore, on the season and time of day. Well-pronounced was photosynthetic increase of pH (=9) in water of L. Hüüdre in 2021 (**Table 1**). Water parameters corresponded to 'poor' or 'bad' ecological status

(**Tables 1** and **3**), and a strong phytoplankton bloom decreased SD to 0.5 m. Submerged vegetation was absent, among floating-leaved plants *Sparganium gramineum* and some common species occurred. Temporary photosynthetic alkalization is among the stressors acting in unbalanced SWL. In **Table 1** become evident, that most high pH values are joint with higher nutrient contents. In the period of the highest nutrient loading (1981 in **Table 1**) L. Solda was measured with the highest pH value: 10. Alkalization during the intensification of photosynthesis is described in the isoetid-lake Uljaste (**Figure 6**), in 1993, where pH raised >9.5 and started fish kill. The reason for fish kill was a complex of phenomena: high water temperature, a wide range of oxygen content fluctuation during a short period, very high pH, water bloom caused by toxin-producing cyanobacteria and ammonia toxification [17].

As mentioned above, elodeids (including potamids) appear in SWL with the increase in trophic level. Eutrophication may be caused by direct nutrient loading but also by increased alkalinity. Water enrichment with minerals in SWL may take place even by adding unwashed sand in a swimming place. Carbonates may come from eroded steep slopes (L. Piigandi; **Table 1**), but in Scandinavia, they were added mainly by liming. According to a profound study by Lucassen et al. [18] on Norwegian carbon-limited Isoëteto-Lobelietum SWL, liming decreased redox potential and increased availability of HCO_3^-, CO_2, NH_4^+, PO_4^{3-} and Fe^{2+} in sediment pore water. In Sweden, the occurrence of several acid-sensitive plants, *Myriophyllum alterniflorum* among them, increased in many limed lakes [19]. Interestingly, Lucassen et al. [18] concluded that *Sparganium angustifolium* is more dependent on the quality of sediment and its pore water than on the nutrients in the water since this bur-reed appeared in the limed plots. According to these authors, it is able to take up CO_2 by roots. Our study of the sediments (0–20 cm) of L. Jussi Pikkjärv, located on the

Figure 6.
Lake quillwort in littoral of L. Uljaste, endangered by opening of new mines (author's photo).

limestone bedrock in North Estonia, revealed that $CaCO_3$ content in dry weight under the stands of *S. angustifolium* formed 0.5–2.7%; organic matter 48–96% [20]. We can suppose that nutrient resource for the fast development of long leaves of *S. angustifolium* is indeed supported by sediment because all trophy parameters of the water were among the lowest of SWL studied in 2021: HCO_3^- 5.3 mg L^{-1}; COD 23 mg L^{-1}; dissolved organic carbon 9.8 mg C L^{-1}; P_{tot} 0.013 mg L^{-1}; N_{tot} 0.54 mg L^{-1}; pH 6.5. In the lake occurred dwarf water lily *Nuphar pumila*; in the shallower part *Isoëtes lacustris*, but abundant was also *Potamogeton natans* L., inhabiting eutrophic-hypertrophic lakes as well. We can conclude that water parameters are not always reflecting the sediment properties of different macrophytes.

Anthropogenic liming in Estonia has been mostly unintended. Some SWL and peat bogs of Northeast Estonia were limed by the oil-shale ash from thermal power stations [21]. In the same region, the drinking water supply lowered the water level of some SWL 3–4 m in the 1970s, resulting in loss of isoetids and growth of pine trees in former littoral. Remarkably, in this period charophytes appeared in the SWL of this region, again disappearing after the raising of the water level. However, changed littoral with remains of trees was no anymore suitable for isoetids. Lowering of the water level combines several effects: alkalization due to increased impact of groundwater (**Table 1**, L. Kurtna Valgejärv in 1981), increased nutrient concentrations, new habitats for emergent belt, promoted by increasing trophy, and advantage of eutraphents among hydrophytes. Nowadays new mines threaten the SWL of Northeast Estonia.

Due to rarity of SWL, limnological studies are concerned mainly with direct nutrient loading, frequently resulting in the mass propagation of phytoplankton or large filamentous algae as suppressors or competitors. All these chemical and biological factors are unfavorable also for SWL, their ecosystem become unbalanced. As seen above, the relations of alkalization and trophy level are more complex than presumed, and water level may modify their impact.

4.3 Competition

For submerged hydrophytes worsening of the light conditions due to decreased water transparency by phytoplankton blooms or thick epiphyton may be decisive. Our experiment with seeds of *S. angustifolium* and *S. gramineum* [9, 20] revealed that there exist unrecognized threats. Besides low germination percentage (varying in sediment samples 2–17%), all seedlings declined under the thick cover of green filamentous algae (*Oedogonium* sp.), despite repeated removal of algal bunches under stereomicroscope during several months. Very probably, algae used decaying seedlings also as a nutrient resource. Supposedly the algae derived from the mud samples that originated from the studied lakes. It can be supposed that also in natural conditions seedlings may perish, but this process on the lake bottom is not observable by a field investigation. In the course of 10 macrophyte investigations of L. Piigandi (in 1953–2022), vegetative plants of *S. gramineum* with floating leaves were found four times and fertile once. In 2 years, only seedlings were found and three times bur-reeds were absent. Perishing of seedlings in L. Piigandi is very probable also under the masses of elodeids that increase their abundance at increased alkalinity. In L. Mähuste thick loose algal mat has been covering the beds of *Isoëtes lacustris* in the depth of 1–2 m in the last years. It is not possible to ascertain the presence of *Isoëtes echinospora*, earlier growing mixed with *I. lacustris*. The duration of this algal mat during the vegetation period is not studied, probably there are periods of lesser abundance. However, photosynthesis of isoetids is hindered under the algal mat. Near the distribution area of isoetids is a holiday place, in use since the Soviet time. According to information from local people, even military tanks were sometimes washed in the lake.

Species growing deeper are more dependent on water transparency, thus, plants with fine-divided leaves, enabling larger surfaces for photosynthesis, are in advantage. It is notable, that the appearance of alternate water milfoil in 14 lakes in South-East Estonia coincided with intense eutrophication and erosion from 1960s to 1980s. During the last 20 years, this species has disappeared from 11, remained in three and inhabits four new lakes. In the lakes where milfoil recently has the highest abundance, the ecological status according to all indicators (**Table 3**) was 'moderate' or 'poor' in the last years. In the lakes filled by *M. alterniflorum*, it displaces other rare species: *Isoëtes lacustris*, *Sparganium gramineum* and *Lobelia dortmanna*. In L. Pullijärv, alternate water-milfoil was found in 1968 and occurs as a dominant until now (**Figure 7**). In 2012 L. Tsolgo Pikkjärv was inhabited by a large single pure population of *S. gramineum*, but water-milfoil, by itself under the cover of algal mass, had destroyed the population of bur-reed in 2021 (**Figure 8**). In the particular lake, our germination experiment [9] ascertained viable seeds of *Sparganium* in the sediments, and the samples contained also spores of *Isoëtes lacustris* and seeds of *Elatine*, growing in the lake before the loading from the pigsty on the bank had its effect. Now their growth zone is covered by an emergent belt. For disturbed SWL *M. alterniflorum* can be considered as eutraphent, despite its rarity. Also, *Myriophyllum sibiricum* Komar., rapidly spreading in very different lakes, has been found in 10 SWL during the last decade.

Applied law does not allow the eradication of one rare species in favor of the other of the same category even in a small part of the lake. In the author's opinion, we must consider the global distribution areas and the level of extinction risk in such cases. The area of *M. alterniflorum* includes North America and Eurasia, but *S. gramineum* is

Figure 7.
In L. Pullijärv water lobelia inhabits depth zone to 1 m. deeper on, to 4 m, grows alternate water-milfoil (author's photo).

Figure 8.
Stands of Sparganium gramineum *in L. Tsolgo Pikkjärv in 2012 (above) and the same area occupied by* Myriophyllum alterniflorum *and algae in 2021 (below; author's photos).*

distributed in North Europe and in Siberia, occurring rarely everywhere. In this context, the attempt to save *S. gramineum* is urgently needed. Of course, we cannot be sure that we have adequate information about survival for suppressing species in all parts of its distribution area. Making a decision may sometimes be complicated due to rapid changes also for species that were more common previously.

The studies by Bertrin [22] in the littoral of the lakes of Southwest France (Lacanau, Casaux-Sanguinet, Perentis-Biskarrosse) found that isoetids tolerate mechanical stress by wind and waves relatively well, and our data affirm this conclusion. In the presence of invasive species *Lagarosiphon major* and *Egeria densa* in the lakes studied by Bertrin and co-authors, isoetids persisted better in the areas of higher wind stress. In Estonia water lobelia, inhabiting zone up to 1 m grows under the wave stress in open littoral or in the narrow, sparse and low reed belt. In general, increase and densification of emergent belt is unfavorable for isoetids; also the tuft sedge may reach 1 m water depth and between the tufts organic sediments accumulate.

4.4 Holiday-makers

After the end of flax retting period 50 years ago, many saunas have been built on the banks of SWL, mostly without water purification. Several smaller SWL have lost their typical vegetation. Nowadays, when the mobility of people has increased, the risk of invasive species introduction or diseases affecting the native biota, has sharply increased.

Unfortunately, the most sensitive OL lakes are under the greatest pressure of holiday-makers, because their banks are sandy within the whole perimeter (**Figure 9**). Especially beloved are larger shallow areas, attracting swimmers. In the 1 m zone plants are trampled out and into pieces—all places of frequent swimming are without isoetids. The same result is reported also from France [22]. Besides destruction of plant beds, rich assortment of chemicals is used for washing, UV-protection, warding off insects, etc.; added are the nutrients by swimmers themselves. The shores of OL lakes are very vulnerable to treading due to poor and thin soil layers and also erosion from sandy banks with scarce vegetation increases remarkably. In SD lakes, the survival of isoetids depends on the proportion of sandy and peaty stretches and the number of holidaymakers. Eutrophied B-lakes are under less pressure because masses of elodeids and wider emergent belts make them less attractive.

5. Conclusions

To the protected hydrophyte taxa of Estonian SWL belong isoetids *Isoëtes lacustris*, *Isoetes echinospora* and *L. dortmanna*; floating-leaved *S. angustifolium* and *S. gramineum*, amphibious *Elatine hydropiper* and elodeid *M. alterniflorum*. Isoetids are characteristic for A-lakes: water alkalinity <30 mg HCO_3^- L^{-1}. Elodeids occur mainly in B—lakes: 30–80 HCO_3^- L^{-1}, and their role increases by eutrophication. The relationship between alkalization and eutrophication is complicated because alkalization enhances nutrient enrichment and eutrophication supports the rise of alkalinity, including higher pH in peak hours of photosynthesis. The potential of sediments for nutrient supply of isoetids and *Sparganium* does not be reflected in water parameters (**Figure 10**).

Anthropogenic acidification of Estonian lakes is not known, but natural dystrophication due to the inflow of humic compounds is probable. In most SWL

Figure 9.
All banks of L. Viitna Pikkjärv are sandy, so entrances of swimmers are apparent in many places — isoetids are lost (author's photo).

eutrophication and alkalization have counterbalanced the dystrophication, but in some lakes with unsettled catchment disappearance of isoetids and *Sparganium* obviously has been caused by this process. Water parameters are not always reflecting the sediment properties for different macrophytes.

Competition between primary producers takes place as an occupation of littoral habitats by widening emergent belts, suppression by elodeids and overshadowing by phytoplankton blooms or macroalgae. Filamentous algae may exterminate slow-growing plants already in seedling stage. Fast-growing eutraphents produce more biomass and the accumulation of organic matter decreases oxygen content of littoral sediments.

Nowadays, the mobility of people has increased, and the risk of invasive species or diseases has sharply increased. The most sensitive OL lakes are under the greatest

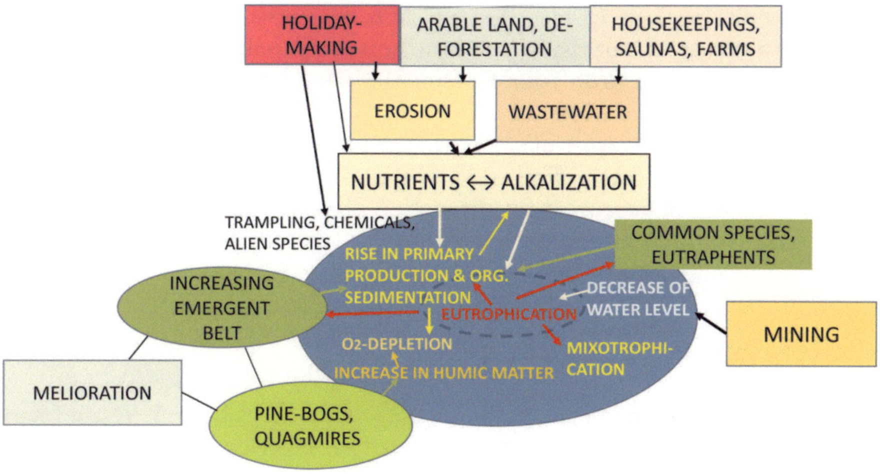

Figure 10.
Scheme of impacts on the soft-water lakes. Very probably, available knowledge is not adequate to depict all possible relationships and interferences.

pressure of holiday-makers, trampling in isoetid zone, causing erosion from lake banks and using chemicals.

Investigations of SWL during the decades have raised the necessity for better organization of the work. Under EU Water Framework Directive carried out monitoring of ecological status mainly on larger, > 10 ha lakes. In future, it is highly recommendable to join hydrochemical and biological monitoring with inventories of rare macrophytes and expand the selection to the smaller SWL. Also, the databases must be joined on behalf of better protection.

In the case of competition between two rare species and the need to save suffering species, it is recommendable to consider the global distribution areas and compare the level of extinction risk of the species.

In principle, it is almost impossible to reverse back the changes in lake ecosystems that formed and balanced over thousands of years and became unbalanced during a much shorter period of anthropogenic intervention. The processes depicted here are complex and interactive, and it would be much easier to keep functioning ecosystems. Theoretically, it would be possible to remove eutraphents, redundant masses of organic sediments and to stop any anthropogenic input. Though the removal of sediments may cause loss of the seed bank, and then the plants must be re-settled from the existing populations if they are locally available. However, our experiment with seeds revealed that there exist unpredictable difficulties.

Main part (about 90%) of SWL with the rare macrophytes are officially protected in Estonia as lakes of type 3110 according to the EU Habitat Directive (*Natura* 2000), and mostly they belong to the larger protected landscape areas. Regrettably, *Natura*-protection is not sufficient for the water bodies, because larger or lesser part of their catchment remains unprotected. Also protection zone 10 m according to the Estonian Water Law is considered mainly by official institutions, private owners are not interested to know and follow the laws and control about their activities is weak. Logging of timber at water border is frequent.

In practice, the protection get tangled already in the first steps—on the level of political authorities, partially due to continuing under-evaluation of the importance of declining habitats and their species. As long as incompetence in ecology is not seen as a shortage for a career, the decisions are based on the populism or business interests. On the other hand, nature protection is shamed by unqualified implantation, by mechanically following of deficient enactments, without clarification and (friendly) discussion with people. The role of scientists in nature protection, theoretically extremely important, could be eliminated by ignoring their advices. The most common practice is using of resources for protection plans (made by scientists) without the accomplishment of recommended activities.

Acknowledgements

Hydrobiological and chemical studies on soft-water lakes since 1992 are financed by Estonian State Monitoring programs; repeated inventories of protected plant species since 2011 by Estonian Environment Agency, Estonian Environmental Board, and in 2021 by Estonian Environment Investment Centre (project 18745). Field works in 2011–2012 were assisted by numerous workers of Estonian Environmental Board and my family members. Since 2018, the studies were carried out with Dr. Kadi-Palmik-Das. I am greatly indebted to my good colleagues: to Dr. Reet Laugaste for determining the algae, compiling a figure and for good advices, and to Prof. Ingmar Ott for generalization of hydrochemical limits of the lake types and revision of the manuscript. An excellent long-lasting cooperation in field, at preparing the reports and compiling the manuscripts with Dr. Kadi Palmik-Das has been pleasant. A good cooperation with Master's Student Eliise Kara at *Sparganium*-investigation in 2021 was interesting and gave a hope that new generation of scientists could continue with nature protection more effectively. All these colleagues have notably improved the manuscript. For this work was used the herbarium of mosses and vascular plants of Estonian University of Life Sciences: TAA.

Abbreviations

SWL	soft-water lake(s)
COD	dichromate oxygen consumption
P_{tot}	total phosphorus
N_{tot}	total nitrogen
OL	oligotrophic
SD	semidystrophic
SD	Secchi depth = water transparency

Author details

Helle Mäemets
Estonian University of Life Sciences, Tartu, Estonia

*Address all correspondence to: helle.maemets@emu.ee

References

[1] Ott I, Kõiv T. Estonian Small Lakes: Special Features and Changes. Tallinn; 1999. p. 128

[2] Webster KE, Bowser CJ, Anderson MP. Understanding the Lake-groundwater system: Just follow the water. In: Magnuson J, Kratz TK, Benson BJ, editors. Long-Term Dynamics of Lakes in the Landscape. Oxford: University Press; 2006. p. 400

[3] Arold I. Eesti maastikud. Tartu: Tartu Ülikooli Kirjastus; 2005. p. 453

[4] Mäemets A, Mäemets A. General characteristics of the lakes. In: Timm T, editor. State of the Estonian Soft-Water Lakes (in Russian, Summary in English). Tartu; 1991. pp. 9-38

[5] Poikane S, editor. Water Framework Directive Intercalibration Technical Report. Institute for Environment and Sustainability; 2009. p. 174

[6] Mäemets A. Macrophytes. In: Timm T, editor. State of the Estonian Soft-Water Lakes (in Russian, Summary in English). Tartu; 1991. pp. 80-122

[7] Paal J, Pärnsalu P-R, Mäemets H. Quagmires around southern and southeastern Estonian lakes. Mires and Peat. 2019;**24**:1-25. DOI: 10.19189/MaP.2019.OMB.StA.1766

[8] Uibopuu I. Eesti pehmeveeliste järvede sammaltaimed [thesis]. Tartu: Tartu University; 2015

[9] Kara E. Seed-Bank of Protected Floating-Leaved Bur-Reeds in Four Soft-Water Lakes in Estonia and Its Implications for Conservation [thesis]. Tartu: Estonian University of Life Sciences; 2022

[10] Murphy KJ. Plant communities and plant diversity in softwater lakes of northern Europe. Aquatic Botany. 2002; **73**(4):287-324. DOI: 10.1016/S0304-3770(02)00028-1

[11] Sand-Jensen K, Borum J, Binzer T. Oxygen stress and reduced growth of Lobelia dortmanna in sandy lake sediments subject to organic enrichment. Freshwater Biology. 2005;**50**:1034-1048. DOI: 10.1111/j.1365-2427.2005.01382.x

[12] Mäemets H. *Sparganium* in Soft-water Lakes of Estonia. 2015. DOI: . Available from: https://www.researchgate.net/publication/281319200_Sparganium_in_soft-water_lakes_of_Estonia

[13] Roll H. Isoetes, Lobelia und Litorella in kalkarmem und kalkreichem Wasser (Ein Beitrag zur Ökologie des Isoeteto-Lobelietum Koch). Beihefte zum Bot. Centralbl. 1939;**Bd. LIX**:345-358

[14] Haab H. Ida-Eesti viieteistkümne toitainevaese ja pehmeveelise järve veesammalde ökoloogia ja levik [thesis]. Tartu: Tartu University; 1987

[15] Order Document of Estonian Minister of Environment 21.04.2020. Pinnaveekogumite nimekiri, pinnaveekogumite ja territoriaalmere seisundiklasside määramise kord, pinnaveekogumite ökoloogiliste seisundiklasside kvaliteedinäitajate väärtused ja pinnaveekogumiga hõlmamata veekogude kvaliteedinäitajate väärtused

[16] Smolders AJP, Lucassen ECHET, Roelofs JGM. The isoetid environment: Biochemistry and threats. Aquatic Botany. 2002;**73**(4):325-350. DOI: 10.1016.S0304-3770(02)00029-3

[17] Ott I. What killed the fish in it? The state of a lake is reflected by its water. Estonian Nature. 1994:53-54

[18] Lucassen ECHET, Spierenburg P, Fraaije RGA, Smolders AJP, Roelofs JGM. Alkalinity generation and sediment CO2 uptake influence establishment of *Sparganium angustifolium* in softwater lakes. Freshwater Biology. 2009;**54**:2300-2314. DOI: 10.1111/j.1365-2427.2009.02264.x

[19] Brandrud TE. Effects of liming on aquatic macrophytes, with emphasis on Scandinavia. Aquatic Botany. 2002;**73**: 395-404. DOI: 10.1016/S0304-3770(02) 00032-3

[20] Mäemets H, Kara E. Lamedalehise ja ujuva jõgitakja seisund Eestis ning soovitusi olukorra parandamiseks. Report for Estonian Environmental Board. 2022

[21] Karofeld E. On technogenic changes in the vegetation and functioning of the bogs of the Kurtna Kames Field. In: Abstracts of the Republican IV Conference in Ecology. Tartu; 1988. pp. 62-64

[22] Bertrin V. Écologie et déterminisme physique des peuplements de macrophytes dans les lacs naturels peu profonds [thesis]. Bordeaux: L'Université de Bordeaux; 2018

Chapter 6

Case Study: Upland Ponds Provide On-Site Mitigation for Bat Habitat along American Electric Power's 765-kV Powerline ROW in the Appalachian Mountains, USA

Virgil Brack Jr, Dale W. Sparks and Scott Kennedy

Abstract

Initially, mitigation for protected species in the United States was project-specific, in-kind, and on-site, benefiting local populations of focal species. Recent mitigation policy uses *in-lieu* fees and mitigation banks, consolidating mitigation at large off-site locations, targeting regional population. This is true for the Indiana bat (*Myotis sodalis*), a species that roosts in trees and forages along many types of woodland edges. Drinking water is essential but considered ubiquitous and rarely factored into mitigation. In high-relief areas, runoff is rapid, precipitation is low, and evapotranspiration is high in late summer, limiting drinking pools for lactating females and juveniles. In Virginia's Ridge and Valley region, our on-site mitigation paired ponds, roost boxes, and edge foraging habitat along a new electric transmission line right-of-way (ROW). During mitigation, white-nose syndrome (a fungal disease) decimated populations of this and four additional species of bats, which we included in analyses. Mitigation metrics were abundance, presence of focal species, species richness, and species diversity. The Indiana bat was known regionally, and an adult male was captured pre-construction along the ROW but not at mitigation sites; the species was recoded acoustically at a mitigation site post-construction. For other focal species, abundance (total and reproductive females), was dramatically greater post-mitigation. Species richness and diversity increased severalfold post-mitigation.

Keywords: on-site mitigation, small-scale mitigation, mitigation success, Indiana bat, northern-long-eared bat, white-nose syndrome, created wetland, utility corridor

1. Introduction

When mitigation is a regulatory requirement, it must comply with laws and implementing regulations. According to the United States (U.S.) President's Council on Environmental Quality (40 CFR [Code of Federal Regulations] Part 1508.20) mitigation for the National Environmental Policy Act (NEPA) occurs when project impacts are offset by any combination of (a) avoidance, (b) minimization; (c) repair, rehabilitation, or restoration, (d) reduction or elimination via preservation and maintenance, and/or (e) compensation by replacing or providing substitute resources or environments. The U.S. Fish and Wildlife Service's (FWS) first mitigation policy (FR [Federal Register] Vol 46 No. 15 pp7644-7763, 23 January 1981) adopted this definition and provided guidance for timing, location, and type of compensatory mitigation. Mitigation was to last for the life of the project and afterwards for as long as the loss persisted (in-time). Mitigation banking was largely considered a matter of timing (i.e., mitigate now for future impacts). The preferred site for mitigation under the 1981 guidance was as close as feasible to the impact area (on-site), prioritized by plan area, proximity, and ecoregion, so that species associated with a habitat loss remained relatively stable over time in or near the impact area. In-kind mitigation was preferred, replacing lost habitat values (the suitability of an area to support a species), with no net loss of value. FWS's 1981 policy did not cover species listed under the Endangered Species Act (ESA). But under ESA section 7, actions of a Federal agency or applicant that could result in incidental take of listed species (without jeopardy to the species or adverse modification of critical habitat) must mitigate (avoid, minimize, and compensate) negative impacts of the project's development [1]. Practitioners are directed to species recovery plans to help identify opportunities for mitigation, largely on-site.

At its simplest, habitat values lost with development are mitigated by areal replacement of habitat of equal quality. However, such replacement is fraught with complications. Preserving existing habitat results in a net loss of habitat so replacement requires habitat creation, but creating a mature habitat may take decades, resulting in a time disparity between impact and replacement. Further, not all habitat, even if suitable, is of the same quality (e.g., pine versus hardwood forests), and the resource (food, water, or shelter) in shortest supply may limit its value. Alternatively, management, restoration, and enhancement may improve the quality of mitigation habitat, contributing to no net loss while lessening areal requirements.

In eastern North America, mitigation for the endangered Indiana (*Myotis sodalis*) and threatened northern long-eared (*M. septentrionalis*) bats, listed under the ESA, has focused on shelter in the form of natural woodland [2, 3], and occasionally artificial [4] roosts. Foraging habitat for these bats is broad-based [5], including areas near and within forests, such as edges and small openings, open woodlands, solitary trees, space above the canopy, and sometimes agricultural lands [6–12]. Nevertheless, foraging habitat may be as essential to maternity colonies as roosts [13]. Fidelity to foraging areas can determine roosts use [14] and limited roost availability may force bats to commute farther to forage [15], impacting reproductive success.

In the eastern U.S., water is typically considered readily available, even ubiquitous, to maternity colonies of the Indiana bat [5]. Its availability is rarely factored into mitigation, although the Indiana bat uses areas with water disproportionate to availability [13, 16], as do congeners [17]; small and ephemeral water bodies may be important habitat components [18]. Carter [19] stated that maternity colonies of the Indiana bat are associated with bottomland, riparian, wetland, or other hydric forest types in the midwestern U.S., whereas in the Appalachian Mountains, where such areas are limited, maternity colonies in uplands are small and ephemeral. The first successful mitigation employing habitat creation for the Indiana bat incorporated wetlands [20]. Congeners roost near water sources [21, 22] and ensuring water availability is standard conservation for many species of temperate bats [23].

Insectivorous bats cannot survive without drinking water [24–27]. Lactating females drink more [28] and use water resources more, even when they are not readily available [29], such as in high relief mountainous habitat [30–34]. In high-relief areas, bats frequent small water pools (even road ruts) in the Appalachians [35] and Ozarks [36] of the U.S., and Slovakia, Europe [37]. Logging and other human land uses have eliminated may natural wetlands, including those in high-elevations [38, 39]. The greatest need for water by lactating females and newly volant young is in late summer, coincident with low precipitation, high evapotranspiration, and low surface water retention in areas of high relief and high altitude [40–42], all of which limit pools of drinking water. See Hoy [43] for a discussion of upland wetland types, hydrology, and seasonality. On narrow high-gradient streams, bank vegetation can further narrow the space over pools, making it difficult for bats to drink [44–46], despite frequent association of bats with larger streams [18].

Based on ESA section 7 consultation for an electric transmission line, we undertook voluntary conservation to improve on-site habitat quality. Along the right-of-way (ROW), we created small ponds to provide drinking water for bats through late summer, and we erected artificial roost boxes adjacent to the ponds. Woodlands edging the ROW provided foraging habitat. Initially, the focus of this mitigation was the Indiana bat, based on local occurrence [6, 47]. During mitigation monitoring, white-nose syndrome (WNS), a fungal disease that kills bats hibernating in caves, decimated population of many species in the eastern U.S. Monitoring analyses were expanded to species of bats that were both (a) considered rare by federal (FWS and Jefferson National Forest [JNF]) and state (Virginia and West Virginia) agencies, and (b) were demonstrably in the area (i.e., focal species): northern long-eared, little brown (*M. lucifugus*), eastern small-footed (*M. leibii*), and tricolored (*Perimyotis subflavus*) bats. Mitigation efficacy was assessed for focal species pre- versus post-project (i.e., pre-construction and post-mitigation) and pre- and post-WNS.

Use of ponds (and boxes) is a form of on-site, small-scale mitigation designed to aid local populations. We did not use in-lieu fee payments or mitigation banks that consolidate mitigation for multiple projects at large off-site locations and often target benefit to regional population. This trend has been increasingly encouraged by FWS in recent years [48]. We selected on-site, small-scale mitigation because (a) trading impacts at one location for mitigation at another is frequently inequitable [49], (b) small-scale mitigation can be very effective [50], (c) combining mitigation construction with project construction (on this and other projects) can be cost-effective, and (d) mitigation banks were not available as recent as a decade ago, long after project initiation [51] (in 2013 the Department of the Interior, Office of Policy Analysis listed 111 banks, but none for bats).

2. Methods

2.1 The project

American Electric Power Service Corporation (AEP) initiated a project in 1998 to construct a 765-kV, 145-km electric transmission line in Appalachian regions of southwest Virginia and West Virginia, U.S. (**Figure 1**). Studies were completed along 55 km of ROW, 17.7 km in JNF in Bland, Wythe, and Pulaski counties, Virginia (**Figure 1**), in the Ridge and Valley physiographic region where elevation is 610–1227 m, local relief

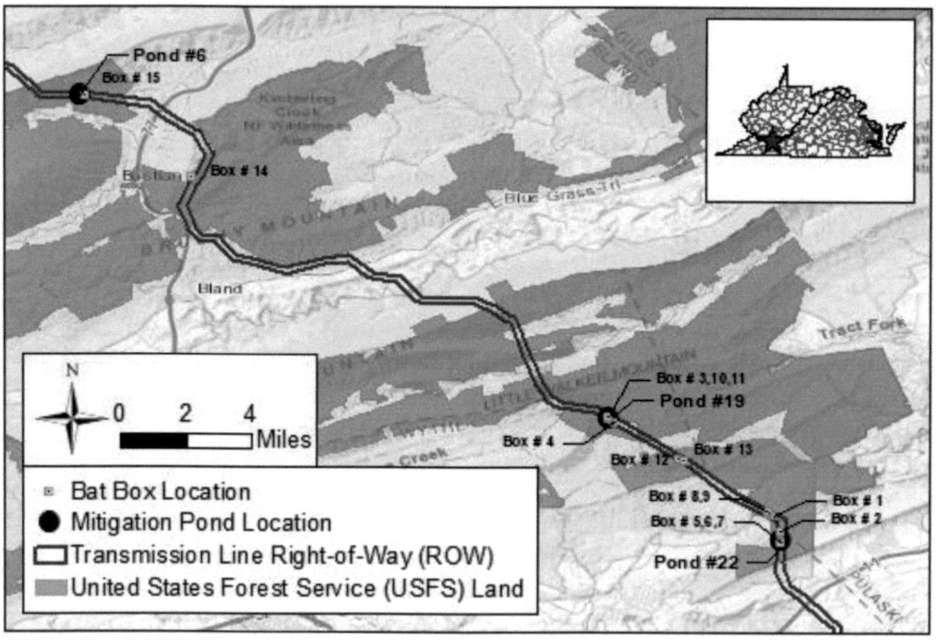

Figure 1.
AEP's 765 kV electric transmission line project in West Virginia and Virginia where mitigation consisting of ponds and boxes was employed.

is >300 m, and mean annual rainfall is 89–102 cm. Ridge tops are underlain by sandstones while mountain bases and valleys have cave-forming limestone and dolomites. It is within the Oak-Chestnut Forest described by Braun [52]. Winter hibernacula include natural caves and anthropogenic mines.

Construction created an open ROW 60 m wide, new and widened, on a high-relief forested landscape. In addition to the ESA, AEP's compliance included numerous federal and state laws and regulatory agencies. One interagency outcome was voluntary conservation under the ESA to mitigate impacts to the Indiana bat. This included constructing upland ponds to provide drinking water (and aquatic insect prey), with adjacent roosting (roost boxes), and foraging (woodlands edging the ROW) habitats, along portions of the ROW crossing JNF (**Figure 1**).

For the 55 km of ROW (including 17.7 km in JNF) addressed by this mitigation, pre-construction bat surveys were completed in summer 2004. Project construction, including ROW clearing, tower and pond construction, and ROW reclamation included the period after netting 2004 through early 2007. Post-construction monitoring was initiated in summer 2007. Pond construction was contained within the ROW and did not constitute any additional impact, areal or temporal, greater than ROW construction impact.

2.2 Bat sampling

Bat surveys pre- and post-construction of the transmission line followed FWS protocol (current version at [53]) that specifies sampling effort, placement of sampling devices, and interpretation of results. Nets, and later acoustic bat detectors, were placed along edges because foraging and commuting bats frequent convex and concave woodland corridors, including utility ROWs, roads, and forest edges [16, 34, 54–59].

Suitable habitat may require the proximity of commuting corridors [18]. Per protocol [53], pre-project mist net capture of a listed bat denotes species presence and is sufficient to require ESA determination on whether development of a project will result in "take" and whether mitigation is required. We used this same presence or probable absence criteria to evaluate results of post-construction surveys, with addition of acoustic detection.

Pre- and post-construction sampling consisted of tending two nets per net site for 5 h/night for two nights (i.e., four net nights [NN]/site) during suitable weather during the summer maternity season (15 May–15 August). Pre-construction netting was completed 15 May–12 July 2004 at 53 sites along the proposed ROW in Virginia and West Virginia. This included the three sites (6, 19, and 22) where ponds were constructed on JNF lands. Post-construction monitoring surveys 2007, 2008, 2009, 2011, and 2013 consisted of netting and bat detector (acoustic) recordings at the three ponds sites.

Bat detectors were not an approved sampling technique during pre-construction sampling but were approved late during post-construction sampling. Bat echolocation calls were collected using AnaBat (Titley Enterprises, LLC) detectors. Per FWS protocol, the sample effort was two detector-nights (DN)/site (pond). Comparisons among sites and years were standardized by limiting data to nights when detectors recorded for 5 h and data after that time frame were excluded.

2.3 Data analyses

Metrics of mitigation success were abundance, presence of focal species, species richness, and species diversity [60]. Mist net effort was standardized per FWS protocol at 4 NN/site for each sample. Abundance was standardized as capture of bats/NN for all species, individual focal species, and reproductive females (pregnant, lactating, and post-lactating females). Project abundance was compared to five published local and regional (hereafter regional) studies (**Table 1**) completed using the FWS protocol. Comparisons among project samples and regional baseline populations (the median of five regional pre-WNS populations) were made as proportions (**Tables 2** and **3**). Detector results were expressed as calls/DN).

Richness was the number of species in any sample, and the diversity index [65] used was diversity = $1/\sum Pi^2$, where Pi is the proportion of bats belonging to species i, often stated as the number of equally represented species. Evenness of capture across space, time, and among species was compared using a Chi-square (χ^2) tests. Acoustic data were analyzed using Kaleidoscope Pro (Version 3.1.1), approved by FWS in 2016. When Kaleidoscope identified calls made by bats of the genus *Myotis* (except the small footed bat), they were also identified visually (vetted) by biologists approved by FWS.

2.4 Mitigation ponds

Pre-construction net sites were assessed for their ability to receive and sustain a small pond providing drinking water through the summer maternity season. Ponds were constructed at net sites 6, 19, and 22 (hereafter pond sites) in late-summer to early-autumn 2006 (**Figure 1**). Ponds were designed to US Forest Service standards, with a surface area of 140 m^2 sloped to a depth of 1–2 m [66] (**Figure 2**). A certified wetlands biologist visited ponds following construction and summers 2007–2010 and 2012 to assure they held water and were functional.

Bat species	Sample	No. NN	Bats/NN	Reproductive female bats/NN
All	2004 ROW	212	1.118	—
	2004 future ponds sites	12	0.083	—
	2007–2009 pond sites pre-WNS	36	4.167	—
	2011–2012 ponds sites post-WNS	24	0.417	—
	WV pre-WNS 1997–2008[1]	3577	3.416	—
	WV post-WNS 2010[1]	892	1.390	—
	WV pre-WNS 2002[2]	63	1.444	—
	WV pre-WNS 1998[3]	176	0.75	—
	VA pre-WNS 2000–2009[4]	804	1.959	—
	VA pre-WNS 1995–1996[5]	24	2.333	—
Indiana	2004 ROW		0.005	0
	2004 future ponds sites		0	0
	2007–2009 pond sites pre-WNS		0	0
	2011–2012 ponds sites post-WNS		0	0
	WV pre-WNS 1997–2008[1]	3577	0.010	0.001
	WV post-WNS 2010[1]	892	0.001	0.001
	VA pre-WNS 2000–2009[4]	804	0.001	0
	Pre-WNS: WV 2002[2]; 1998[3]; VA 1995–1996[5]	63; 176; 24	0	0
N. long-eared	2004 ROW		0.557	0.302
	2004 future ponds sites		0.083	0.083
	2007–2009 pond sites pre-WNS		0.889	0.278
	2011–2012 ponds sites post-WNS		0	0
	WV pre-WNS 1997–2008[1]	3577	1.438	0.583
	WV post-WNS 2010[1]	892	0.330	0.126
	WV pre-WNS 2002[2]	63	0.746	0.349
	WV pre-WNS 1998[3]	176	0.216	0.125
	VA pre-WNS 2000–2009[4]	804	0.412	0.170

Bat species	Sample	No. NN	Bats/NN	Reproductive female bats/NN
	VA pre-WNS 1995–1996[5]	24	0	0
Little brown	2004 ROW		0.198	0.085
	2004 future ponds sites		0	0
	2007–2009 pond sites pre-WNS		0.472	0.083
	2011–2012 ponds sites post-WNS		0	0
	WV pre-WNS 1997–2008[1]	3577	0.836	0.247
	WV post-WNS 2010[1]	892	0.168	0.048
	WV pre-WNS 2002[2]	63	0.111	0.048
	WV pre-WNS 1998[3]	176	0.125	0.023
	VA pre-WNS 2000–2009[4]	804	0.313	0.113
	VA pre-WNS 1995–1996[5]	24	0	0
E. small-footed	2004 ROW		0.009	0.009
	2004 future ponds sites		0	0
	2007–2009 pond sites pre-WNS		0	0
	2011–2012 ponds sites post-WNS		0	0
	WV pre-WNS 1997–2008[1]	3577	0.042	0.016
	WV post-WNS 2010[1]	892	0.007	0.003
	WV pre-WNS 2002[2]	63	0.016	0
	VA pre-WNS 2000–2009[4]	804	0.095	0.034
	Pre-WNS: WV 1998[3]; VA 1995–1996[5]	176; 24	0	0
Tricolor	2004 ROW		0.042	0.005
	2004 future ponds sites		0	0
	2007–2009 pond sites pre-WNS		0.194	0
	2011–2012 ponds sites post-WNS		0	0
	WV pre-WNS 1997–2008[1]	3577	0.215	0.010
	WV post-WNS 2010[1]	892	0.049	0.008
	WV pre-WNS 2002[2]	63	0.048	0.016

Bat species	Sample	No. NN	Bats/NN	Reproductive female bats/NN
	WV pre-WNS 1998[3]	176	0.091	0.006
	VA pre-WNS 2000–2009[4]	804	0.246	0.040
	VA pre-WNS 1995–1996[5]	24	0.083	0.042

[1][61]: WV, 37 counties (31 pre-WNS; 8 post-WNS).
[2][47]: Preston County, WV.
[3][62]: Randolph County, WV
[4][63]: 11 counties, Cumberland Plateau and Ridge and Valley physiographic provinces, VA.
[5][64]: southeast VA.

Table 1.
Abundance, bats/per NN, including reproductive females for all bats and focal species during the 2004 preconstruction survey at all 53 net sites and at 3 pond sites (6, 19 and 22), post construction 2004–2009 (pre-WNS) and 2011–2013 (post-WNS), and for 5 regional studies pre- and one post-WNS.

Bat species	ROW pre-construction pre-WNS		Regional pre-WNS	Reproductive female pre-WNS:ROW pre-construction vs regional
	vs regional pre-WNS	vs regional post-WNS	vs regional post-WNS	
All	0.537	0.804	1.499	—
Indiana	1.000	5.000	5.000	0.000
N. long-eared	0.775	1.688	2.179	1.036
Little brown	0.474	1.179	2.488	0.689
E. small-footed	0.189	1.286	6.786	0.189
Tricolor	0.286	0.857	3.102	0.210

Table 2.
*Proportional abundance (bats/NN) at 53 ROW sites, pre-construction and pre-WNS, with median values of 5 regional studies (see **Table 1**), pre- and post-WNS, between regional pre- and post-WNS values, and between reproductive females (pregnant, lactating, and postlactating) pre-WNS—ROW and regional values.*

(a) (b)

Figure 2.
Pond site 22 construction (a) within and coincident with ROW development in August 2006, and (b) this same pond after first filling October 2006.

	Pond sites pre-construction		Pond sites post-construction		
Bat species	**vs ROW**	**vs regional pre-WNS**	**vs pond sites pre-construction**	**vs ROW**	**vs regional pre-WNS**
All	0.074	0.040	50.205	3.727	2.000
Indiana	0.000	0.000	0.000	0.000	0.000
N. long-eared	0.149	0.115	10.711	1.596	1.236
Little brown	0.000	0.000	Infinite	2.384	1.129
E. small-footed	0.000	0.000	0.000	0.000	0.000
Tricolor	0.000	0.000	Infinite	4.619	1.320

Table 3.
A comparison (proportion) of abundance (bats/NN) at 3 pond sites pre-construction, post-construction (pre-WNS), and post-construction and post-WNS, with the 53 sites on the ROW, and the median value of 5 local and regional projects pre-WNS (see **Table 1** *and foot-note) a single regional project post-WNS.*

2.5 Management adaptations: technology, science, and disease

Successful mitigation requires the ability to respond to unforeseen circumstances. Over the more than 2 decades of this project, adaptive management has increasingly become a requirement for ESA mitigation [67] but was rarely implemented when this project was initiated [68–71]. Over 20 years there were many changes in politics and regulatory policy, technology, scientific understanding, and disease that affected mitigation implementation and outcome, and also project finances. Many changes were made, and funds shifted to accommodate unforeseen challenges and opportunities.

At project initiation, pre-construction sampling, and mitigation design bat detectors were not part of FWS sampling protocol. However, technology and regulatory acceptance evolved rapidly so we incorporated this technology in mitigation monitoring. This adds complexity because acoustic detection was not available to assess impacts but was available to assess mitigation success. FWS's protocol for acoustics use included guidance for field use (detector placement, level of effort, and acceptable environmental conditions), interpretation of recordings (computer-assisted and human vetting), and interpretation of an acoustic "capture" (like a mist-net capture). This protocol changed over time. After monitoring was complete, we reassessed data using the approach outlined in the current (2022) sampling protocol to ensure consistency.

The original mitigation design included installation and monitoring of 30 metal-clad boxes at 15 locations with solar exposure along the ROW in JNF. This included one location near pond site 6, two near pond site 19, and three near pond site 22 (**Figure 1**). Boxes had 7 (58 × 68.5 × 18 cm) or 14 (58 × 68.5 × 38 cm) chambers; 10 locations had two 7-chamber boxes and 5 locations had one 7-chamber and one 14-chamber box. Two boxes were mounted back-to-back 3 m high on metal poles (**Figure 3**) facing roughly southeast and northwest to provide a diversity of temperatures inside boxes over a range of environmental conditions. This design allowed assessment of box use in relation to temperatures inside and out, elevation, orientation, and potentially pond use. Boxes were equipped with Thermocron iButton data loggers (Maxim Integrated Products, Sunnyvale, California). These were removed

Figure 3.
Small and large (7 and 24 chamber) metal-clad bat boxes erected on metal poles within the transmission line ROW.

after one season when we learned they emit high frequency sounds that may disturb bats [72]. Boxes were monitored for bats annually, and nests of stinging insects were removed during winters.

In 2011 wooden boxes on wooden poles were added near pond sites 6 and 19 (two each) because of concerns that bats might receive a nuisance static-electrical shock when landing on the metal-clad boxes that go to ground and so deter use. During initial monitoring, no bats were observed in boxes, but guano was collected at metal clad boxes at sites 7 (April 2010) and 8 (June 2013) and in 2010 under a wooden box (site 11.1). In 2021, 1 big brown bat (*Eptesicus fuscus*) was in a box at site 5, and guano was under six metal-clad boxes (sites 1, 2, 5, 6, 7, 15) and one wooden box (15.1.). These limited data do not support the hypothesis that metal-clad boxes on metal poles deterred bat use. At mitigation design, use of environmental DNA, including feces to identify species was in its infancy. In 2012, having lost roost box thermal data and with the advent of WNS in the project region [61], we added guano screens below boxes to collect feces for DNA testing to improve detection of use. However, DNA in feces degrades rapidly and we were unable to identify species.

Vernal pools and small wetlands are important for many species [73]. We designed ponds as functioning wetlands although they were not used for wetland mitigation credits. We expected to gain a natural complement of obligate and facultative wetland plants and animals through natural processes and species complements were monitored. Control of non-native invasive plants was not planned but they and aggressive native cattails (*Typha* sp.) were removed by hand as time allowed. Although fenced, cattle were allowed into one pond on two occasions; they were removed, and the fence repaired.

The biggest impact to the original mitigation plan and assessing success, was the arrival of WNS. We expanded 5 years of annual monitoring to bi-annually for the last two surveys because WNS demonstrably affected summer bat populations [61] and expanded analyses of mitigation success to include pre- and post-WNS effects.

3. Results

3.1 Pre-construction

In total, 237 bats representing nine species were captured during the 2004 pre-construction survey at 53 sites. The capture rate was 1.118 bats/NN (**Table 1**) and mean richness was 1.7 species/site. Capture included an adult male Indiana bat, 118 northern long-eared bats (64 reproductive females) at 33 sites, 42 little brown bats (18 reproductive females) at 14 sites, 2 reproductive female eastern small-footed bats at 2 sites, 9 tricolored bats (1 reproductive female) at 7 sites, 43 big brown bats, 13 eastern red bats (*Lasiurus borealis*), 7 silver-haired bats (*Lasionycteris noctivagans*), and 2 hoary bats (*Lasiurus cinereus*). Catch was not evenly distributed across species (χ^2 = 440.6582, P < 0.001) and collective species diversity was 3.5.

Abundance of all species combined from the ROW was 53.7% of the median regional value (**Table 2**). Similarly, abundance of focal species was typically less than median regional values, ranging from 18.9% for the eastern small-footed bat to 100% for the Indiana bat (**Table 2**). Abundance of reproductive female northern long-eared bats from the ROW was 103.6% of regional, but for other focal species it was 0–68.9% of regional values (**Table 2**).

At the three sites where ponds were later constructed, only one bat, a pregnant northern long-eared bat, was captured during the pre-construction survey. The catch was 0.083 bats/NN, species richness was 1.0, and diversity was 1.0. Abundance of species collectively and individual focal species was less than along the ROW (0–14.9%) and median regional values (0–11.5%) (**Table 3**).

3.2 Post-construction

Sampling at ponds produced 160 bats representing 7 species over 5 monitoring seasons, 2007–2013: 49 (31%) big brown, 53 (33%) eastern red, 32 (20%) northern long-eared, 16 (10%) little brown, 7 (4%) tricolored, 2 hoary, and 1 silver-haired bats. All bats of focal species were caught pre-WNS. At pond sites 6, 19, and 22 respectively, 32, 48, and 80 bats were caught. Captures differed among ponds ($\chi^2 = 22.4$; $P < 0.001$), years ($\chi^2 = 110.1$; $P < 0.001$), and species ($\chi^2 = 126.3$; $P < 0.001$). More bats were typically caught at pond sites 22 than 19 and 6. Ninety-four percent of captures (150 of 160) were pre-WNS. Post-mitigation abundance, richness, and diversity increased pre-WNS and decreased post-WNS at each pond site (**Figure 4**) and collectively (**Table 1**).

Abundance at pond sites of combined species post-construction and pre-WNS was >5000% of pre-construction pond sites, 373% of pre-construction ROW, and 200% of regional values (**Table 3**). Similarly, abundance of focal species (Indiana, northern long-eared, little brown, eastern small-footed, and tricolored bats) at pond sites post-construction and pre-WNS was 112.9–1071% compared to pre-construction of ponds, ROW, and regional values (**Table 3**). Abundance of reproductive females followed a similar pattern (**Table 1**). Abundance of all species at pond sites post-construction and post-WNS was 502% of preconstruction, 10%, of post-construction and pre-WNS, 20% of regional values pre-WNS, and 30% of regional values post-WNS.

Abundance based on calls/DN at pond sites post-construction, was greater pre- than post-WNS for all species combined (189.4 versus 51.9 calls/DN), and for Indiana (1.5 versus 0.3), northern long-eared (1.9 versus 0.1), little brown (20.1 versus 0.6), and tricolor (80.4 versus 4.8) bats. Calls for the eastern small footed bat were not vetted.

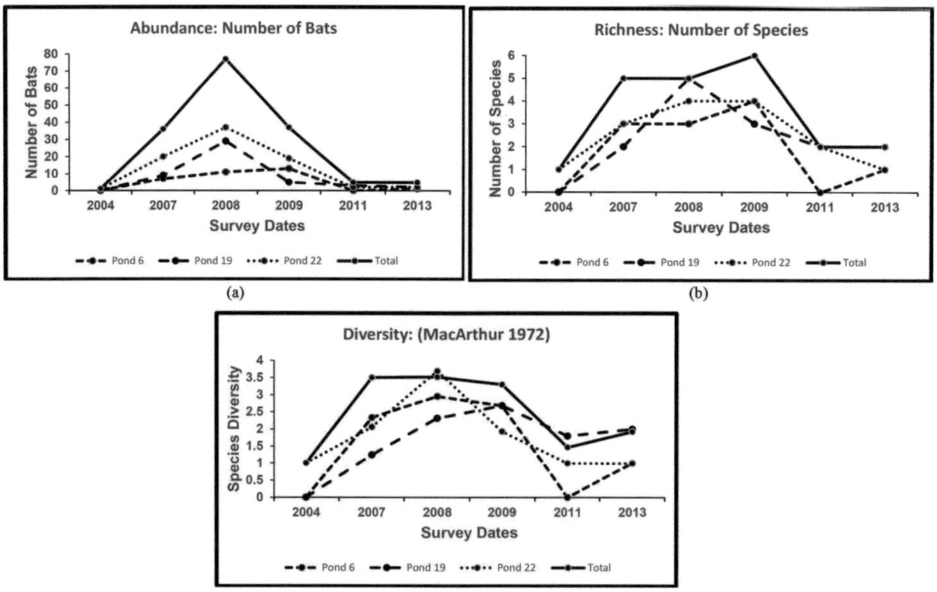

Figure 4.
Abundance (a), richness (b), and diversity (c) of bat captures at each of 3 pond sites including the pre-construction survey (2004) and 5 post-construction monitoring surveys pre-WNS (2007–2009) and post-WNS (2011 and 20013).

4. Discussion and conclusions

Low bat abundance, especially reproduction females, at high altitudes [30, 34, 74–76] is often attributed to cool variable temperatures, but drinking water is also less common in areas of high-relief, frequent at high altitudes [77]. If water is limited in late summer when need is greatest, lactating females and young bats travel farther to drink, potentially limiting juvenile recruitment [78]. In late summer on mountainous landscapes, drinking water can be more limiting than prey [33]. Providing drinking water, in conjunction with corridors used for commuting and foraging [8, 11, 56], may benefit bats more than replacing tree roosts on a heavily wooded landscape. At a minimum, bats use and benefit from natural and created pools across a broad geography and habitat type [79–84].

4.1 Mitigation success pre-WNS

Rare populations are hard to sample [60, 85–87]. For regulatory compliance under ESA, capture of just one bat, via mist net or acoustic detection, is demonstrable proof of presence. In this project area of steep relief and high altitude, for the initial focal species, the Indiana bat, success was limited. Although not netted at pond mitigation sites before or after mitigation, a single individual was captured along the project ROW pre-project, which represented a higher level of abundance than regional studies. This species was also documented acoustically at a pond during mitigation, and per FWS protocol, documents presence and thus limited success. This scenario also played out for the eastern small-footed bat, although acoustic recordings were not considered. For the three remaining focal species and for all species combined, abundance (measured by netting and acoustics) was dramatically greater after mitigation than pre-project at (1) the mitigation site, (2) the project ROW, and (3) regionally, whereas pre-project abundance at mitigation sites was lower than regional values. Abundance of reproductive females of focal species was also greater after mitigation. Richness and diversity increased severalfold with pond mitigation.

While improved abundance (notably of reproductive females), richness, and diversity support success of pond mitigation, these are not the only criteria of success. From a regulatory perspective, rare or unique species may be valued more than abundance, but for this project abundance of all species and focal species improved with mitigation. Similarly, diversity is not a stand-alone assessment criterion. For example, abundance may be low, and desirable species may be absent or fail to reproduce despite high diversity. But that did not happen with this mitigation. Similarly, species richness may be devalued if desired species are absent or invasive species abound, but that also was not true. In addition, maintaining or enhancing species richness and diversity may be important when combating threats like disease, e.g., WNS [88], and development, e.g., wind energy [89]. Richness and diversity on a broad scale may provide resilience against local stochastic events that jeopardize small and isolated populations.

4.2 Mitigation success post-WNS

WNS threatens extinction or regional extirpation of many bats, including focal species of this study [88]. Despite abundant documentation within hibernacula [90], this was one of the first documented effects of WNS on summer populations [82, 91–95]. Based on netting and acoustic sampling, presence of listed and focal species, abundance at multiple levels, species richness, and diversity decreased dramatically with onset of WNS. Post-WNS captures were limited to eastern red and hoary bats, species for which WNS mortality is not documented, and the big brown bat, a species that suffers limited effects from WNS [96]. Nevertheless, post-WNS, mitigation exhibited limited success based on collective abundance of all species, richness, and diversity.

4.3 On-site mitigation

Mitigation can be implemented locally on-site where impacts occur, or at a distant off-site location, often at a large scale [48] (typically a bank) and in recent years funded by *in-lieu* fee payments. Both on-site, often small-scale, and offsite large-scale mitigation have advantages and disadvantages. Large scale mitigation is a recent trend [48], particularly for bats. Frequently-cited benefits are consolidating mitigation for many small, isolated impacts, especially in areas with high development pressure, while lessening collective administrative burdens and potentially costs [97, 98]. However, evidence of claimed benefits of large-scale mitigation is generally lacking [99], and few studies determine overall effect on individuals or populations [100]. Small-scale mitigation is just sometimes better [50].

Replacing lost habitat is often the first mitigation choice. Replacing lost woodlands at a distant location arguably does not benefit individuals directly affected, a goal of ESA mitigation, and preserving existing habitat at a distant site does not support a no-loss or net gain objective (FR Vol. 81, No. 248, pp. 95316-95349). Wild populations are strongly stochastic [101, 102] so maintaining or improving a broad local population [50] provides resiliency against stochastic events [103–105]. This is especially important for long-lived species [106] such as bats. Long-term conservation of many species of bats may depend on maintaining local viable populations across a wide geographic area. Replacing lost habitat locally is not always possible because of competing land uses. A heavily wooded landscape (such as Appalachia) may offer limited opportunities and diminishing returns for reforestation. The timeline for creating a forest, 50–80 years [107, 108], is inconsistent with timing of the impact [109]. Appalachia mine lands may provide a mitigation opportunity, but mining typically impacts woodlands so reforestation of mined lands where forest habitat was lost does before provide no net loss, mitigation timelines may be even longer [110], and mining and has its own mitigation requirements (Surface Mining Control and Reclamation Act or SMCRA) that may favor competing uses that conflict with bat mitigation [111].

Alternatives to habitat creation, on- or off-site, are habitat restoration or enhancement. Standing water is a resource required and frequently used by bats [79–84, 112]. Standing water is uncommon because in late summer in areas of high relief and altitude wetlands dry seasonally and during drought [113, 114], and many wetlands have been lost because of past land uses [38, 39]. Wetland enhancement and restoration improve habitat and benefit bats by providing drinking water, which is the goal of mitigation.

Effective mitigation is in-time, off-setting impacts at the time they occur. Constructing mitigation ponds during project construction does this; bats begin to use ponds as soon as they are available [84]. Further, regulatory compliance forces construction to be completed off-season (non-reproductive season) when bats are at hibernacula, making ponds available for filling with spring rains before bats return for the summer reproductive season. Combining project and mitigation construction saves money and it increases the probability mitigation is completed. Mitigation that is complex and time consuming for project proponents and regulatory agencies is often ineffective or is not completed [115, 116]. Combining mitigation and project construction spatially and temporally creates no additional adverse impacts and reduces costs. This mitigation is also a cost savings over the long-term capital investment to grow large trees on an additional mitigation site. In short, providing drinking water is not just ecologically sound, but can be cost-effective, a stated goal of endangered species conservation [117].

Simple, cost-effective, successful mitigation should be given priority. Regulatory compliance policy streamlining or providing approval, *a priori*, for situations where ponds (and boxes) can be used for mitigation, would benefit applicants and agencies while producing good outcomes. Prime candidates for such approval are utility ROWs crossing forested landscapes. Although typically considered an impact rather than an asset, ROWs are a conservation resource for many species. This sometimes includes rare species [118], and many species of bats uses these concave corridors [18, 56, 119]. Mitigation based on banking and related *in-lieu* fee programs are limited by the geographic availability of operational banks because of land availability, direct and administrative costs, and the time to create functional banks. Ponds (and boxes) placed near foraging habitat provide functional, small-scale, on-site, cost effective mitigation.

Acknowledgements

We are grateful to many individuals who participated in a project spanning 20 years. AEP funded the studies; we note R Poff, J Van Hassel, and J Henry. JNF encouraging mitigation within their boundaries; we note K Landgraf, S Croy, and J Overcash. L Robbins identified species from acoustic surveys. T Brack, then ESI's CFO and now CEO, consistently allowing the project to go beyond bounds of profitability and funded manuscript preparation.

Abbreviations

AEP	American Electric Power Service Corporation—one of the largest generators of electricity in the U.S., with more than 5.5 million customers in 11 states, and with 40,000 miles of lines has the largest transportation network in the U.S.
CFR	Code of Federal Regulations—official legal publication used by departments and agencies of the U.S. Federal Government to codify rule making
ESA	Endangered Species Act—enacted 28 December 1973, it establishes protections for fish, wildlife, and plants listed as threatened or endangered and provides for interagency cooperation to avoid take of listed species
FR	Federal Register—official daily publication of the U.S. Federal Government for rules, proposed rules, and notices of Federal agencies and organizations, and executive orders and other presidential documents
FWS	U.S. Fish and Wildlife Service—a bureau in the U.S. Department of the Interior, with primary responsibility for conservation and management of fish, wildlife, plants, and their habitats

JNF	Jefferson National Forest—a U.S. National Forest in southwest Virginia, administered with George Washington National Forest, collectively one of the largest blocks of public land in the eastern U.S.
NEPA	National Environmental Policy Act—U.S. law signed 1 January 1970, which requires federal agencies to assess and disclose environmental effects of proposed actions prior to decision-making
NOAA	National Oceanic and Atmospheric Administration—a regulatory agency in the U.S. Department of Commerce that monitors oceanic and atmospheric conditions and forecasts weather
ROW	right-of-way—the easement corridor on which a utility, road, or railroad line is built and operated.
SMCRA	Surface Mining Control and Reclamation Act - passed 20 May 1977, is the primary U.S. federal law regulating environmental effects of coal mining
U.S.	United States of America
WNS	white-nose syndrome - a fungal disease that kills hibernating bats, which has decimated population of many species in the eastern U.S.

A. Appendix

If water is limiting, bats likely use ponds more during drought, influencing interpretation of post-construction sampling. However, as long-term practitioners in the project area's high elevation, high-relive topography, we know that net sites spaced by 1 km, as defined by FWS sampling protocol, often experience very different

Year	Pond	July	August
2007	6	75–110	10–75
	19	75–150	25–50
	22	50–90	50–100
2008	6	75–110	50–90
	19	90–150	100–150
	22	90–110	110–150
2009	6	50–100	75–110
	19	50–90	75–110
	22	75–125	75–100
2011	6	25–75	25–75
	19	25–90	25–75
	22	50–150	25–90
2013	6	90–125	50–110
	19	150–300	90–125
	22	200–300	25–90

Table A1.
Range of percent of normal precipitations in July and August within 4 km of the three pond sites (6, 19, and 22) in JNF during each of the 5 post-construction sample years.

precipitation on any night, as dramatically as no rainfall to torrential downpours (**Table A1**). Thus, localized precipitation can produce a spotty distribution of available drinking water, and effects of drought, across the landscape.

FWS protocol defines an active (foraging) area for the Indiana bat as a circle with a 4 km radius (2.5 mi), or 5,082.6 ha (50.8 km^2), so we used this to define the area within which drought might determine the availability of drinking water.

The National Oceanic and Atmospheric Administration (NOAA) [120] provides precipitation on a 4 by 4 km (16 km^2) grid. We examined late summer monthly precipitation, July and August, as a percent of normal at the three pond locations in JNF during each of the five post-construction sample years. We examined range of percent of normal precipitation for 4 by 4 km grids within 4 km of pond sites (**Table A1**).

If we define a local drought, within 4 km of a pond, as <90% of expected precipitation throughout, then pond 6 experienced droughts in July and August in 2011 whereas ponds 6 and 19 experienced August droughts in 2007 and 2011. If we define a drought as <75% of expected precipitation throughout, then only pond 19 in August 2007 experienced drought. In contrast. Precipitation was >150% only in 2013 at ponds 19 and 22.

Post-construction, bat capture differed among ponds (χ^2 = 22.4; P < 0.001) and years (χ^2 = 110.1; P < 0.001). Clearly, WNS was the largest contributor to a difference among years and we cannot ascertain impacts of drought; 94% of captures were pre-WNS (2007–2009). More bats were typically caught at pond 22 than 19 and 6, which again provides no clear effect of drought, particularly pre-WNS.

Regardless of local weather conditions, every pond held water at every visit post-construction, including summer 2020, 15 years after construction. As such, even during local drought conditions, these ponds provided bats drinking water (and potentially an aquatic-insect food-base). While these observations do not help provide a graded metric of the value the ponds provide, they do help demonstrate that the mitigation goal of providing drinking water to focal species of bats during late summer was met.

Author details

Virgil Brack Jr[1]*, Dale W. Sparks[1] and Scott Kennedy[2]

1 Environmental Solutions and Innovations, Inc., Applied Research, Cincinnati, OH, USA

2 American Electric Power Service Corporation, Transmission Line Siting, Roanoke, VA, USA

*Address all correspondence to: vbrack@envsi.com

References

[1] USFWS & NMFS. Endangered Species Consultation Handbook [Internet]. 1993 Draft, 1998 Final. Available from: Endangered Species Consultation Handbook | FWS.gov. [Accessed 11 October 2022]

[2] USFWS. Conservation Strategy for Forest-dwelling Bats in Tennessee [Internet]. 2015. Available from: https://www.fws.gov/cookeville/pdfs/conservation_strategy_for_forest_dwelling_bats_in_Tennessee_10_16_15.pdf. [Accessed 27 October 2021]

[3] USFWS. Revised Conservation Strategy for Forest Dwelling Bats in the Commonwealth of Kentucky [Internet]. 2016. Available from: https://www.fws.gov/frankfort/pdf/20160601%20Revised%20FDBatConservStrategy.pdf. [Accessed 27 October 2021]

[4] Whitaker JO Jr, Sparks DW, Brack V Jr. Use of artificial roost structures by bats at the Indianapolis International Airport. Environmental Management. 2006;**38**:28-36

[5] USFWS. Indiana bat (*Myotis sodalis*) Draft Recovery plan: First Revision [Internet]. 2007. Available from: https://www.fws.gov/midwest/endangered/mammals/inba/pdf/inba_fnldrftrecpln_apr07.pdf. [27 October Accessed 2021]

[6] Brack V Jr. Autumn activities of the endangered Indiana bat (*Myotis sodalis*) near a hibernaculum in Bland County, Virginia. Northeastern Naturalist. 2006;**13**:421-434

[7] Brooks JD, Loebb SC, Gerard PD. Effect of forest opening characteristics, prey abundance, and environmental factors on bat activity in the Southern Appalachians. Forest Ecology and Management. 2017;**400**:19-27

[8] Grindal SD, Brigham RM. Impacts of forest harvesting on habitat use by foraging insectivorous bats at different spatial scales. Ecoscience. 1999;**6**:25-34

[9] Muller J, Brandl R, Buchner J, Pretzsch H, Seifert S, Stratz C, et al. From ground to above canopy—bat activity in mature forests is driven by vegetation density and height. Forest Ecology and Management. 2013;**306**:179-184

[10] Patriquin KJ, Barclay RMR. Foraging by bats in cleared, thinned and unharvested boreal forest. Journal Applied Ecology. 2003;**40**:646-657

[11] Tibbels AE, Kurta A. Bat activity is low in thinned and unthinned stands of red pine. Canadian Journal Forest Research. 2003;**33**:2436-2442

[12] Wolcott KA, Vulinec K. Bat activity at woodland/farmland interfaces in central Delaware. Northeastern Naturalist. 2012;**19**:87-98

[13] Divoll TJ, Aldrich SP, Haulton GS, O'Keefe JM. Endangered Myotis bats forage in regeneration openings in a managed forest. Forest Ecology and Management. 2022;**503**:119757. DOI: 10.1016/j.foreco.2021119757

[14] Slough BG, Jung TS. Little brown bats utilize multiple maternity roosts within foraging areas: implications for identifying summer habitat. Journal of Fish and Wildlife Management. 2020;**11**:311-320

[15] Bender MJ, Castleberry SB, Miller DA, Wigley TB. Site occupancy of foraging bats on landscapes of managed pine forest. Forest Ecology and Management. 2015;**336**:1-10

[16] Kniowski AB, Gehrt SD. Home range and habitat selection of the Indiana bat in an agricultural landscape. Journal Wildlife Management. 2014;**78**:503-512

[17] Krusic RA, Yamasaki M, Neefus CD, Pekins PJ. Bat habitat use in White Mountain National Forest. Journal of Wildlife Management. 1996;**60**:625-631

[18] Gallagher ME, Farrell SL, Germain RH, Rojas VG. Summer bat habitat use and forest characteristics in managed Northeastern forests. Journal of Forestry. 2021;**119**:305-318

[19] Carter C. Indiana bats in the Midwest: The importance of hydric habitats. Journal Wildlife Management. 2010;**70**:1185-1190

[20] Sparks DW, Brack V Jr, Whitaker JO Jr, Lotspeich R. Reconciliation ecology and the Indiana bat at Indianapolis International Airport. In: Larauge PB, Castille ME, editors. Airports: Performance, Risks, and Problems: Hauppauge. New York: Nova Science Publishers; 2009. pp. 51-65

[21] Johnson JS, Kiser JD, Watrous KS, Peterson TS. Day-roosts of *Myotis leibii* in the Appalachian Ridge and Valley of West Virginia. Northeastern Naturalist. 2011;**18**:95-106

[22] Kalcounis-Rueppell MC, Psyllakis JM, Brigham RM. Tree roost selection by bats: an empirical synthesis using meta-analysis. Wildlife Society Bulletin. 2005;**33**:1123-1132

[23] Kalda O, Kalda R, Liira J. Multi-scale ecology of insectivorous bats in agricultural landscapes. Agriculture, Ecosystems and Environment. 2014;**199**:105-113

[24] Kurta A, Bell GP, Nagy KA, Kunz TH. Water balance of free-ranging little brown bats (*Myotis lucifugus*) during pregnancy and lactation. Canadian Journal Zoology. 1989;**67**:2468-2472

[25] Kurta A, Kunz TH, Nagy KA. Energetics and water flux of free-ranging big brown bats (*Eptesicus fuscus*) during pregnancy and lactation. Journal of Mammalogy. 1990;**71**:59-65

[26] McLean JA, Speakman JR. Energy budgets of lactating and non-reproductive brown long-eared bats (*Plecotus auritus*) suggest females use compensation in lactation. Functional Ecology. 1999;**13**:360-372

[27] Webb PI, Speakman JR, Racey PA. Defecation, apparent absorption efficiency, and the importance of water obtained in the food for water balance in captive brown long-eared (*Plecotus auritus*) and Daubenton's (*Myotis daubentoni*) bats. Journal of Zoology. 1993;**230**:619-628

[28] Adams RA, Hayes MA. Water availability and successful lactation by bats as related to climate change in arid regions of western North America. Journal Animal Ecology. 2008;**77**:1115-1121

[29] Amorim F, Jorge I, Beja P, Rebelo H. Following the water? Landscape-scale temporal changes in bat spatial distribution in relation to Mediterranean summer drought. Ecology and Evolution. 2018;**8**:5801-5814

[30] Grindal SD, Morissette JL, Brigham RM. Concentration of bat activity in riparian habitats over an elevational gradient. Canadian Journal Zoology. 1999;**77**:972-977

[31] Hintze F, Duro V, Carlos J, Carvalho A, Eira C, Rodrigues PC, et al. Influence of reservoirs created by

small dams on the activity of bats. Acta Chiropterologica. 2016;**18**:395-408

[32] Lison F, Calvo FF. Bat activity over small ponds in dry Mediterranean forests: Implications for conservation. Acta Chiropterologica. 2014;**16**:95-101

[33] Seibold S, Buchner J, Bassler C, Muller J. Ponds in acidic mountains are more important for bats in providing drinking water than insect prey. Journal of Zoology. 2013;**209**:302-308. DOI: 10.1111/jzo.12041

[34] Law B, Anderson J, Chidel M. A bat survey in state forests on the south-west slopes region of New South Wales with suggestions of improvements for future surveys. Australian Zoologist. 1998;**30**:467-479

[35] Kiser JD, MacGregor JR. Indiana bat (*Myotis sodalis*) mist net surveys for coal mining activities. In: Vories KC, Harrington A, editors. Proceedings of Indiana Bat & Coal mining: a technical Interactive Forum. Alton, Illinois: DOI Office of Surface Mining; 2004. pp. 45-62

[36] Wilhide JD, Harvey MJ, MacDaniel VR, Hoffman VE. Highland pond utilization by bats in the Ozark National Forest. Arkansas. Arkansas Academy of Science. 1998;**52**:110-112

[37] Kanuch P, Danko S, Celuch M, Kristin A, Pjencak P, Matis S, et al. Relating bat species presence to habitat features in natural forests of Slovakia (Central Europe). Mammalian Biology. 2008;**73**:147-155

[38] Francl KE, Ford WM, Castleberry SB. Characterization of high elevation Central Appalachian wetlands. USDA FS Northeastern Research Station Research Paper. 2004;**NE-725**:1-26

[39] Weakley AS, Schafale MP. Non-alluvial wetlands of the Southern Blue Ridge: diversity in a threatened ecosystem. Water, Air, and Soil Pollution. 1994;**77**:359-383

[40] McGuire KJ, McDonnell JJ, Weiler M, Kendall C, McGlynn BL, Welker JM, et al. The role of topography on catchment-scale water residence time. Water Resources Research. 2005;**41**:1-41

[41] Webster JR, Wallace JB, Benfield EF. Organic processes in streams of the eastern United States. In: Cushing CE, Cummins KW, Minshall GW, editors. Ecosystems of the World 22. New York: Elsevier; 1995. pp. 117-187

[42] Wolman MG, Riggs HC, editors. Surface Water Hydrology: Geology of North America. Vol. 0-1. Boulder: Geological Society of America; 1990

[43] Hoy C. A hydrologic characterization of three headwater mountain wetlands in eastern Kentucky [thesis]. University of Kentucky, Lexington; 2012

[44] Jackrel SL, Matlack RS. Influence of surface area, water level and adjacent vegetation on bat use of artificial water sources. American Midland Naturalist. 2010;**164**:74-79

[45] Korine C, Adams R, Russo D, Fisher-Phelps M, Jacobs D. Bats and water: anthropogenic alterations threaten global bat populations. In: Voigt C, Kingston T, editors. Bats in the Anthropocene: Conservation of Bats in a Changing World. New York: Springer; 2016. pp. 215-241

[46] Razgour O, Korine C, Saltz D. Pond characteristics as determinants of species diversity and community composition in desert bats. Animal Conservation. 2010;**371**:1-9

[47] Brack V Jr, Orndoroff W, Reynolds RJ, Zokaites J, Zokaites C. Bats

of Skydusky Hollow, Bland County, Virginia. Virginia Journal Science. 2005;**56**:93-106

[48] Clement JP, Alletta d'A B, Bean MJ, Boling TA, Lyons JR. A Strategy for Improving the Mitigation Policies and Practices of the Department of the Interior. Washington, D.C.: Report to Secretary of Interior from the Energy and Climate Change Task Force; 2014. p. 25

[49] Ruhl JB, Salzman J. Apples for oranges. Environmental Law Reporter. 2001;**31**:11438-11474

[50] Clevenger AP. Mitigating continental-scale bottlenecks: how small-scale highway mitigation has large-scale impacts. Ecological Restoration. 2012;**30**:300-307

[51] DOI Office of Policy Analysis. Conservation Banking Overview and Suggested Areas for Future Analysis [Internet]. 2013. Available from: https://www.conservationfund.org/images/cln_events-resources/CB_Webinar_Series_2020/Webinar-1/Resources/DOI_Office_of_Policy_Analysis--_CB_Overview_DOI-Sept2013.pdf. [Accessed 11 October 2022]

[52] Braun EL. Deciduous Forests of Eastern North America. Philadelphia: Blackiston; 1950

[53] USFWS. Range-wide Indiana Bat Survey Guidelines. [Internet]. 2020. Available from: https://www.fws.gov/midwest/endangered/mammals/inba/surveys/pdf/FINAL%20Range-wide%20IBat%20Survey%20Guidelines%203.23.20.pdf. [Accessed 27 October 2021]

[54] Kalcounis-Rueppell MC, Briones KM, Homyack JA, Petric R, Marshall MM, Miller DA. Hard forest edges act as conduits, not filters, for bats. Journal Wildlife Management. 2013;**37**:571-576

[55] Law B, Chidel M. Tracks and riparian zones facilitate the use of Australian regrowth forest by insectivorous bats. Journal Applied Ecology. 2002;**39**:605-617

[56] Lesinski G, Olszewski A, Popczyk B. Forest roads used by commuting and foraging bats in edge and interior zones. Polish Journal of Ecology. 2011;**59**:611-617

[57] Lundy M, Montgomery I. Summer habitat associations of bats between riparian landscapes and within riparian areas. European Journal Wildlife Research. 2010;**56**:385-339

[58] Morris AD, Miller DA, Kalcounis-Rueppell MC. Use of forest edges by bats in a managed pine forest landscape. Journal Wildlife Management. 2010;**74**:26-34

[59] Murray SW, Kurta A. Nocturnal activity of the endangered Indiana bat (*Myotis sodalis*). Journal of Zoology. 2006;**262**:197-206

[60] MacKenzie DI, Nichols JD, Sutton N, Kawanishi K, Bailey LL. Improving inferences in population studies of rare species that are detected imperfectly. Ecology. 2005;**86**:1101-1113

[61] Francl KE, Ford WM, Sparks DW, Brack V Jr. Capture and reproductive trends of summer bat communities in West Virginia: assessing the impact of white nose syndrome. Journal Fish and Wildlife Management. 2012;**3**:33-42

[62] Menzel MA, Owen SF, Ford WM, Edwards JW, Wood PB, Chapman BR, et al. Roost tree selection by northern long-eared bat (*Myotis septentrionalis*) maternity colonies in an industrial forest

of the central Appalachian Mountains. Forest Ecology and Management. 2002;**155**:107-114

[63] Timpone J, Francl K, Sparks DW, Brack V Jr, Beverly J. Bats of the Cumberland Plateau and Ridge and Valley Provinces, Virginia. Southeastern Naturalist. 2011;**10**:515-528

[64] Hobson CS. Bat records from southeastern Virginia including a new resident species, *Myotis austroriparius* (Chiroptera: Vespertilionidae). Banisteria. 1998;**12**:18-23

[65] MacArthur RH. Geographical Ecology. New York: Harper and Row; 1972. p. 288

[66] Biebighauser TR. A Guide to Creating Vernal Ponds: All the Information You Need to Build and Maintain an Ephemeral Wetland. Morehead, Kentucky: U.S. Forest Service; 2003. p. 32

[67] Williams BK, Brown ED. Technical challenges in the application of adaptive management. Biological Conservation. 2016;**195**:255-263

[68] Fischman RL, Ruhl JB. Judging adaptive management practices of U.S. agencies. Conservation Biology. 2015;**30**:268-275

[69] Ruhl JB. Taking adaptive management seriously: a case study of the Endangered Species Act. University of Kansas Law Review. 2003-2004;**52**:1249-1284

[70] Runge MC. Adaptive management for threatened and endangered species. Journal of Fish and Wildlife Management. 2011;**2**:220-233

[71] Williams BK, Szaro RC, Shapiro CD. Adaptive Management: the U.S. Department of the Interior Technical Guide. Edition—updated. ed. Washington, D.C.: Department of the Interior; 2009. p. 72. DOI: gov/sites/doi.gov/files/migrated/ppa/upload/TechGuide.pdf

[72] Willis CKR, Jameson JW, Faure PA, Boyles JG, Brack V Jr, Cervone TH. Thermocron iButton and iBBat temperature dataloggers emit ultrasound. Journal of Comparative Physiology B. 2009;**179**:867-874

[73] Kissel AM, Halabisky M, Scherer RD, Ryan E, Hansen EC. Expanding wetland hydroperiod data via satellite imagery for ecological applications. Frontiers in Ecology and the Environment. 2020;**18**:432-438

[74] Brack V Jr, Stihler CW, Reynolds RJ, Butchkoski CM, Hobson CS. Effect of climate and elevation on distribution and abundance in the mid-eastern United States. In: Kurta A, Kennedy J, editors. The Indiana Bat: Biology and Management of an Endangered Species. Austin: Bat Conservation International; 2002. pp. 21-28

[75] Cryan PM, Bogan MA, Altenbach JS. Effect of elevation on distribution of female bats in the Black Hills, South Dakota. Journal of Mammalogy. 2000;**81**:719-725

[76] Ford WM, Menzel MA, Menzel JA, Welch DJ. Influence of summer temperature on sex ratios in eastern red bats (*Lasiurus borealis*). American Midland Naturalist. 2001;**147**:179-184

[77] Kanuch P, Kristin A. Altitudinal distribution of bats in the Poľana Mts area (Central Slovakia). Biologia, Bratislava. 2006;**61**:605-610

[78] Hoying KM, Kunz TH. Variation in size at birth and post-natal growth in the

insectivorous bat *Pipistrellus subflavus* (Chiroptera: Vespertilionidae). Journal of Zoology (London). 1998;**245**:15-27

[79] Bartonicka T, Rehak Z, Andreas M. Diet composition and foraging activity of *Pipistrellus pygmaeus* in a floodplain forest. Biologia. 2008;**63**:1-7

[80] Francl KE. Summer bat activity at woodland seasonal pools in the Northern Great Lakes region. Wetlands. 2008;**28**:117-124

[81] Huie KM. Use of constructed woodland ponds by bats in the Daniel Boone National Forest [thesis]. Eastern Kentucky University; 2002

[82] Johnson JB, Rodrigue JL, Ford MW. Nightly and Yearly Bat Activity Before and After White-nose Syndrome on the Fernow Experimental Forest in West Virginia. Newtown Square, Pennsylvania: U.S. Forest Service; 2013. p. 22

[83] Maslonek ML. Bats use of created and natural wetlands [thesis]. West Virginia University; 2009

[84] Menzel JM, Menzel MA, Kilgoi JC, Ford WM, Edwards JW. Bat response to Carolina Bays and wetland restoration in the southeastern U.S. Coastal Plain. Wetlands. 2005;**25**:542-550

[85] Kalton G, Anderson DW. Sampling rare populations. Journal Royal Statistical Society, Series A. 1986;**149**:65-82

[86] McDonald LL. Sampling rare populations. In: Thompson WL, editor. Sampling Rare or Elusive Species. Washington, D.C.: Island Press; 2004. pp. 11-42

[87] Waples RS. Definition and estimation of effective population size in the conservation of endangered species. In: Beissinger SR, McCullough DR, editors. Population Viability Analysis: Assessing Models for Recovering Endangered Species. Chicago: University of Chicago; 2002. pp. 147-168

[88] Thogmartin WE, Sanders-Reed CA, Szymanski JA, McKann PC, Pruitt L, King RA, et al. White-nose syndrome is likely to extirpate the endangered Indiana bat over large parts of its range. Biological Conservation. 2013;**160**:162-172

[89] Frick WF, Baerwald EF, Pollock JF, Barclay RMR, Szymanski JA, Weller TJ, et al. Fatalities at wind turbines may threaten population viability of a migratory bat. Biological Conservation. 2017;**209**:172-177

[90] Thogmartin WE, King RA, Szymanski JA, Pruitt L. Space-time models for a panzootic in bats, with a focus on the endangered Indiana bat. Journal Wildlife Diseases. 2012;**48**:876-887

[91] D'Acunto LE. Potential influence of white-nose syndrome on summer bat distribution in Pennsylvania [thesis]. Indiana University of Pennsylvania; 2012

[92] Dzal Y, McGuire LP, Veselka N, Fenton MB. Going, going, gone: the impact of white-nose syndrome on the summer activity of the little brown bat (*Myotis lucifugus*). Biology Letters. 2010. DOI: 10.1098/rsbl.2010.0859

[93] Moosman JP, Veilleux J, Pelton JG, Thomas H. Changes in capture rates in a community of bats in New Hampshire during the progression of white-nose syndrome. Northeastern Naturalist. 2013;**20**:552-558

[94] Reynolds RJ, Powers KE, Orndorff W, Ford MW, Hobson CS. Changes in rates of capture and demographics of *Myotis*

septentrionalis (northern long-eared bat) in western Virginia before and after onset of white-nose syndrome. Northeastern Naturalist. 2016;**23**:195-204

[95] Powers KE, Reynolds RJ, Orndorff W, Ford MW, Hobson CS. Post-white-nose syndrome trends in Virginia's cave bats, 2008-2013. Journal of Ecology and Natural Environment. 2015;**7**:113-123

[96] Cheng TL et al. The scope and severity of white-nose syndrome on hibernating bats in North America. Conservation Biology. 2021. DOI: 10.1111/cobi.13739

[97] Hayes DJ. Addressing the environmental impacts of large infrastructure projects: making "mitigation" matter. Environmental Law Report. 2014;**44**:10016-10021

[98] Kennedy CA, Miteva DA, Baumgarten L, Hawthorn PL, Sochi K, Polasky S, et al. Bigger is better: improved nature conservation and economic returns from landscape-level mitigation. Science Advances. 2016;**2**:e1501021. DOI: 10.1126/sciadv.1501021

[99] Grimm M, Koppel J, Geißler G. A shift towards landscape-scale approaches in compensation - suitable mechanisms and open questions. Impact Assessment and Project Appraisal. 2019;**37**:491-502

[100] Cowan MA, Callan MN, Watson MJ, Watson DM, Doherty TS, Michael DR, et al. Artificial refuges for wildlife conservation: What is the state of the science? Biological Reviews. 2021;**96**:2736-2754. DOI: 10.1111/brv.12776

[101] Boyce MS. Population growth with stochastic fluctuations in the life table. Theoretical Population Biology. 1977;**12**:366-373

[102] Higgins K, Hastings A, Sarvela JN, Botsford LW. Stochastic dynamics and deterministic skeletons: population behavior of Dungeness crab. Science. 1997;**276**:1431-1435

[103] Atmar W, Patterson DB. The measure of order and disorder in the distribution of species in fragmented habitat. Oecologia. 1993;**96**:373-382

[104] Wang Y, Chen C, Millien V. A global synthesis of the small-island effect in habitat islands. Proceedings Royal Society B. 2018;**285**:20181868. DOI: 10.1098/rspb.2018.1868

[105] Wright SJ, Hubbell SP. Stochastic extinction and reserve size: a focal species approach. Oikos. 1983;**41**:466-476

[106] Nur N, Bradley RW, Lee DE, Warzybokand P, Jahncke J. Projecting long-term impacts of a mortality event on vertebrates: incorporating stochasticity in population assessment. Ecosphere. 2021;**12**:e03293. DOI: 10.1002/ecs2.3293

[107] Steiner KC, Finley JC, Gould PJ, Fei S, McDill M. Oak regeneration guidelines for the Central Appalachians. Northern Journal of Applied Forestry. 2008;**25**:5-16

[108] Vickers LA, Fox TR, Loftis DL, Boucugnani DA. Predicting forest regeneration in the Central Appalachians using the REGEN Expert System. Journal of Sustainable Forestry. 2011;**30**:790-822

[109] Brack V Jr, Sparks DW. Avoiding a conservation pitfall: considering the risks of unsuitably hot bat boxes: Adding context to Crawford & O'Keefe. Conservation Science and Practice. 2021;**3**:e563. DOI: 10.1111/csp2.563

[110] Skousen S, Ziemkiewicz P, Venable C. Tree recruitment and growth on 20-year-old, unreclaimed surface

mined lands in West Virginia. International Journal of Mining, Reclamation and Environment. 2006;**20**:142-154

[111] Skousen J, Zipper CE. Post-mining policies and practices in the Eastern USA coal region. International Journal of Coal Science and Technology. 2014;**1**:135-151

[112] Brooks RT, Ford MW. Bat activity in a forest landscape of central Massachusetts. Northeastern Naturalist. 2005;**12**:447-462

[113] Moorhead KK. Effects of drought on the water-table dynamics of a southern Appalachian Mountain floodplain and associated fen. Wetlands. 2003;**23**:792-799

[114] Thompson Y, Sandefur BC, Miller JO, Karathanasis AD. Hydrologic and edaphic characteristics of three mountain wetlands in southeastern Kentucky. USA. Wetlands. 2007;**27**:174-188

[115] Gamarra MJC, Toombs TP. Thirty years of species conservation banking in the U.S.: Comparing policy to practice. Biological Conservation. 2017;**214**:6-12

[116] Owley J. Keeping track of conservation. Ecology Law Quarterly. 2015;**42**(79):1-138

[117] USFWS. DOI Memorandum: Guidance for the Establishment, Use, and Operation of Conservation Banks. [Internet]. 2003. Available from: https://www.fws.gov/endangered/esa-library/pdf/Conservation_Banking_Guidance.pdf. [Accessed: 27 October 2021]

[118] Gardiner MM, Riley CB, Bommarco R, Ockinger E. Rights-of-way: a potential conservation resource. Frontiers in Ecology and Environment. 2018;**16**:149-158. DOI: 10.1002/fee.1778

[119] Perry RW, Thill RE, Leslie DM. Scale-dependent effects of landscape structure and composition on diurnal roost selection by forest bats. Journal of Wildlife Management. 2008;**72**:913-925

[120] NOAA. National Weather Services: Advanced Hydrologic Prediction Services [Internet]. 2013. Available from: AHPS Precipitation Analysis (weather.gov). [Accessed: 11 November 2022]

Chapter 7

Giants on an Island: Threats and Conservation Challenges of Elephants Due to Herbivorous Diets

Chandima Fernando and Duminda S.B. Dissanayake

Abstract

Elephants are highly generalized herbivores with a wide dietary range encompassing natural vegetation and cultivated crops. Their foraging strategies vary across different temporal and spatial contexts, as well as among distinct social groups. A significant number of elephants in Asia and Africa reside beyond the boundaries of national parks, nature reserves, and protected areas. Consequently, many elephants face elevated risks of mortality or injury while seeking essential nutrients. This chapter provides an overview of the critical role played by dry-zone forests as habitats for elephants. Furthermore, it explores how human-dominated landscapes influence elephant feeding behaviors and foraging strategies, emphasizing the need to enhance our current understanding of these behaviors and their implications for the future.

Keywords: elephants, foraging strategies, crop-raiding, human-elephant conflict, elephant conservation

1. Introduction

Herbivores play vital roles in maintaining the ecosystem functions and vegetation community through various processes such as grazing, browsing, trampling, and defecation [1–3]. The large (>5 kg) terrestrial mammalian herbivores are considered 'keystone' species in many ecosystems that help define an entire ecosystem. Therefore, without keystone herbivores, the ecosystem would be dramatically different, cease to exist altogether, or could collapse [2, 4]. Therefore, studying many fields of biology, ecology, and conservation of large herbivores has been an essential topic debated by environmental conservationists, ecologists, and management authorities for over a century in many parts of the world. However, current data show that large herbivores face considerable population declines due to various reasons and home range contractions, such that ~60% are threatened with extinction [5].

In the realm of large herbivore research, South and Southeast Asia hold a unique position due to the Indo-Malayan biogeographic realm supporting more than twice the number of large herbivore species per unit area compared to Africa [6]. However, despite this abundance, our understanding of Asian large mammalian herbivores' behavior, population dynamics, community ecology, and ecosystem impacts remains inadequate for effective conservation and management when compared to other regions such as North America, Europe, and Africa [6, 7]. Notably, significant research efforts have been dedicated to understanding the Asian elephant (*Elephas maximus*) and its contribution to the conservation and preservation of its habitats [8].

As the largest living herbivorous animal in Asia, the Asian elephant is distributed across South and Southeast Asia, inhabiting diverse ecosystems, including grasslands, tropical and subtropical forests, marshes, and human-dominated landscapes. However, over the latter half of the twentieth century, the Asian elephant population in the wild has undergone a significant decline [9, 10], resulting in its current endangered status [11] and listing in Appendix I of the Convention on International Trade in Endangered Species of Wild Fauna and Flora (CITES). The remaining wild population of Asian elephants is estimated to be less than 50,000 individuals [8, 10, 12, 13], with three recognized subspecies: *E. maximus indicus* on the mainland, *E. maximus sumatranus* in Sumatra, and *E. maximus maximus* in Sri Lanka, which holds particular importance for elephant conservation efforts [14].

The Sri Lankan elephant (*E. maximus maximus*), referred to as the "Sri Lankan elephant" or simply "elephant" henceforth, holds immense symbolic, cultural, and economic value within Sri Lanka (**Figure 1**). This country boasts one of the highest densities of wild Sri Lankan elephants in South Asia, with a remaining wildland area of 4102 km^2 dedicated to their habitat [15]. However, the rapid growth of the human population, the expansion of agricultural lands, along with the implementation of development projects, especially in the dry zone, have led to a significant decline in elephant populations compared to historical levels. This decline has been exacerbated by the adoption of agro-civilization practices, which involve the use of irrigation systems with reservoirs and canals. The estimated population, once reaching 12,000 individuals [16], has now dwindled to 7000 elephants across the entire island [17].

The progressive encroachment of human-dominated landscapes into the shrinking habitats of Sri Lankan elephants has brought about a pressing issue: the human-elephant

Figure 1.
The Sri Lankan elephant walking on a tank bed at Minneriya National Park Sri Lanka.

conflict (HEC) [18]. This conflict arises from competing demands for space, time, and resources, leading to dire consequences for both elephants and humans, including loss of life. As such, the HEC has emerged as a critical socio-economic and political challenge within Sri Lanka [19–21], posing a significant threat to local biodiversity conservation efforts. The gravity of the situation becomes apparent when considering the staggering number of HEC incidents recorded between 2010 and 2019, reaching 14,516 in total [22]. Tragically, these conflicts have resulted in the death of 2631 elephants during the same period, with potentially numerous unreported casualties due to inadequate incident management [22]. Furthermore, the human death rate has experienced a 50% increase in the past two decades [20, 23], culminating in 121 human fatalities in 2019 alone [22]. Managing this immense conflict has become a primary objective for elephant conservation, and Sri Lanka, as an island nation, holds a distinct position in the conservation and management of these magnificent, large herbivores.

Crop-raiding by elephants is a significant factor contributing to human-elephant conflict (HEC) in Sri Lanka, leading to considerable losses in agricultural produce during individual raiding incidents. This issue carries profound implications for elephant conservation, directly impacting the physical and psychological well-being of farmers, their economic stability, food security, and various social and cultural aspects within Sri Lanka [21, 24–26].

Research conducted by De Silva [27] and Jayawardene [28] has shed light on the annual average crop and property damage caused by wild elephants, ranging between Rs. 6000 and Rs. 30,000 per farming family. These cumulative damages amount to approximately Rs. 560.71 million per cropping season or an average of Rs. 1121.42 million per year, encompassing the entire elephant habitat in Sri Lanka [29]. However, it is important to consider that this estimate may lean toward the higher end, as it might include a significant number of farms that experience substantial elephant-related damage, thereby introducing an upward bias. Consequently, comprehending the general feeding behavior of elephants, their herbivorous dietary patterns, and the resulting HEC from crop-raiding becomes crucial in evaluating its impact on conservation efforts. Addressing these challenges is essential for effective conservation and management of elephants in Sri Lanka.

Exploring the intricate feeding behavior of large herbivores on an island is a vast and challenging topic that cannot be adequately covered in a single chapter. Thus, in the subsequent sections, we provide a general background on the significant role played by dry-zone forests in providing a home for elephants (Section 2). Following this, in Section 3, we delve into the interaction between megaherbivores and human landscapes while also discussing the broader aspects of man-made feeding grounds for elephants. Section 4 focuses on the specific issue of crop-raiding behavior by elephants, shedding light on its implications for the HEC. Additionally, in Section 5, we explore the impact of garbage dump sites on elephants' feeding behavior. Finally, in Section 6, we investigate how the herbivorous diet of elephants influences their habitats. This chapter concludes by presenting emerging threats to elephants, providing a comprehensive overview of the various factors affecting their conservation and management.

2. The dry zone: dry evergreen forests are home to most of the Island's elephants

The dry evergreen forests in the dry zone of Sri Lanka, covering nearly two-thirds of the island, provide crucial habitat for the majority of the country's elephant population.

With 76% of these forests located within the dry zone, they serve as a reliable food source for approximately 5879 elephants [30]. Particularly during times of reduced grass availability, these forests play a significant role in sustaining the elephants [31, 32]. A study conducted in the north-western region of Sri Lanka revealed that elephants consume a diverse array of plant species, including both trees (47%) and non-tree species like shrubs, herbs (including grass), and climbers (53%) [33].

Due to their large size, elephants have substantial dietary requirements, consuming around 150 kg of food daily [13]. The availability of food within the dry zone forests plays a critical role in shaping the home range and movement patterns of Sri Lankan elephants. Sri Lankan male elephants exhibit home ranges that span from 53.6 to 346 km^2, while females have home ranges ranging from 29.6 to 160.7 km^2 [34]. Research by Kumar et al. [35] indicates that elephants exhibit transient foraging behavior, rarely staying in one area for an extended period. Instead, they constantly move between different locations to fulfill their dietary needs. These movement patterns vary throughout the seasons, and their food selection also changes accordingly as they are mixed feeders.

The presence of elephants in dry evergreen forests holds immense ecological significance, as emphasized by the scientific literature [36, 37]. Elephants contribute to the structure and composition of these forests through their feeding habits. By consuming a variety of plant species, elephants facilitate seed dispersal, which promotes forest regeneration and the growth of diverse vegetation [38–42]. Furthermore, their browsing behavior affects the understory vegetation, stimulating the growth of grasses while reducing the dominance of woody plants. Additionally, elephants act as ecosystem engineers by modifying their habitat through activities such as uprooting trees and creating water holes, which benefit other wildlife species [43]. These interactions between elephants and their environment have far-reaching effects on forest dynamics and contribute to the conservation of biodiversity.

Research has demonstrated that the decline of elephants in dry forests can have profound, cascading effects on the ecosystem [44]. Reductions in elephant populations disrupt seed dispersal patterns, leading to diminished tree regeneration and alterations in vegetation structure. Without the influence of elephants' browsing and trampling, the understory vegetation may become denser, impacting the abundance and diversity of plant species. Moreover, the absence of elephants as ecosystem engineers hamper critical ecological processes, including the formation of water sources and the creation of microhabitats for other animals. The overall health and resilience of the dry forest ecosystem are intricately linked to the presence and conservation of elephants.

The dry zone of Sri Lanka has witnessed significant forest losses in recent decades due to various human activities such as resettlements, agricultural encroachments, infrastructure development, and the construction of reservoirs and dams. While these endeavors aim to benefit local communities, they have unfortunately led to the fragmentation of wildlife habitats. As a result, a considerable portion of elephant home ranges now overlaps with human settlements, agricultural lands, and croplands. The construction of development projects that disrupt elephant corridors has further exacerbated the human-elephant conflict (HEC) on the island. The loss of natural habitat not only impacts wildlife but also gives rise to numerous human-wildlife conflicts, including the HEC. Therefore, it is imperative to prioritize the preservation and conservation of the existing dry zone forest cover to protect the natural habitat of Sri Lankan elephants. This necessitates comprehensive land use planning and political stability to ensure the long-term coexistence of elephants and humans on the island.

In order to maintain the ecological integrity and long-term viability of the dry evergreen forests, the conservation and protection of the Sri Lankan elephant population in the dry zone are of the utmost importance. Effective conservation strategies should not only focus on mitigating human-elephant conflicts but also prioritize the preservation of suitable habitats and corridors that facilitate elephant movement. It is crucial to implement measures that reduce habitat fragmentation and secure protected areas within the dry zone. One approach is the establishment of wildlife corridors, which would connect fragmented forest patches and enable elephants to traverse their natural ranges without frequent encounters with human settlements and agricultural activities. By creating and reconnecting these corridors, we can effectively address the challenges posed by habitat fragmentation and promote the coexistence of elephants and humans in the region. Such initiatives should be supported by robust monitoring systems, stakeholder engagement, and the involvement of local communities to ensure the success of conservation efforts in the dry zone of Sri Lanka.

3. Megaherbivore in human landscapes

Over the past few decades in Sri Lanka, rapid spatial and temporal changes in land use by human activities have affected natural landscape structure, patterns, and dynamics. In the island's central dry zone, natural landscape has changed over since the last 2000 years mainly due to man-made tanks and their cascade system, chena cultivation, croplands, and agriculture. These activities reduce the availability of elephants' natural habitats, impacting their movement patterns and home ranges. The cultivated fields with various crops within the home ranges of elephants attract the elephants.

3.1 Tank beds

The dry zone of Sri Lanka is characterized by the presence of numerous man-made lakes known as Wewa or tanks. These reservoirs were constructed by the ancient kings of Sri Lanka to store rainwater and surface runoff during the monsoonal periods [45]. Ranging from small seasonal ponds to vast perennial reservoirs spanning thousands of hectares, these tanks serve as favored habitats for elephants, especially during the dry season [46]. They offer vital water sources to elephants when the water level recedes, revealing fresh green short grass beds that elephants prefer grazing on, as opposed to long grasses.

In the dry zone where elephants reside, it is common to witness elephants congregating in these tank beds in the afternoon, ranging from small family groups to large herds. Throughout the day, elephants are typically found in the *Megathyrsus* grasslands [47]. However, during the evening, they migrate to the seasonal grasslands found on the tank beds. One remarkable gathering of elephants, known as the "greater elephant gathering," takes place in Minneriya-Kaudulla National Park (**Figure 2**). In locations like Wasgamuwa, Sri Lanka, elephants transition to these tank beds at around 15:00 hrs from their daytime refuge, where they begin feeding on short grasses. Elephants rely on bulk feeding of short grasses and a diverse diet to meet their nutritional needs. It is noteworthy that the dietary patterns of Sri Lankan elephants, similar to African elephants, do not solely reflect the availability of grass in tank beds but also involve browsing trees and shrubs in forests [47, 48].

Figure 2.
The Sri Lankan elephant gathering at Minneriya National Park.

These observations highlight the importance of tank beds as a crucial resource for elephants during the dry season. Preserving and conserving these habitats is essential for the well-being of Sri Lankan elephants and for maintaining their dietary diversity. By ensuring the protection of the tank beds and their associated grasslands, we can support the natural foraging behaviors and nutritional requirements of elephants in the dry zone.

3.2 Chena

Chena, also known as slash-and-burn cultivation, holds a significant historical presence in Sri Lanka, dating back as far as 5000 years. It was a prevalent practice among rural villages and early civilizations. To maintain the integrity of the habitat and environment, early Sri Lankans cleared forests specifically for chena practices. The process involved clearing a portion of the forest, cultivating it for 3–4 years, and then abandoning it to establish a new chena elsewhere. Most chena practices were situated near or within natural forests.

The old and abandoned chena lands serve as ideal foraging grounds for elephants [13, 24]. After the chena harvest, a substantial amount of leftover crops remains on the land, attracting elephants in large numbers. According to our observations and conversations with chena farmers, elephants often spend a day or two consuming the leftover crops before moving on to other areas for foraging. Elephants particularly favor excess crops such as millet and beans.

Over time, these abandoned chena lands become colonized by pioneer vegetation, creating rich food sources that attract elephants back to these areas. This behavior confirms that elephants are "edge species" and exhibit a preference for secondary forests [49]. We witnessed this foraging behavior of collard elephants in Lungamwehera, Southern Sri Lanka, as they are annually attracted to these secondary regeneration areas. As the succession progresses, these habitats provide both food and daytime refuge for elephants. In addition to chena, other local farming activities within human-dominated landscapes also create ideal or sub-optimal elephant habitats [13].

In the mid-1900s, the traditional chena practice underwent significant changes due to rapid development, increasing human population, and changing needs. This shift resulted in an environmentally unfriendly cultivation method as chena started to rely on permanent cultivation lands near natural forests in the dry zone of Sri Lanka [23].

The landscape of the dry zone in Sri Lanka underwent dramatic modifications during the Mahaweli Development Project, an irrigation-based agricultural development program initiated in 1978. This project diverted the longest river, Mahaweli, to irrigate the dry-zone landscapes [50]. Consequently, the landscape transitioned into monoculture plantations, croplands, agricultural areas, and developed regions, greatly reducing available habitats for elephants.

3.3 Cropland and agricultural land

Crop and agricultural lands account for approximately one-third of Sri Lanka's total land area. Numerous studies have demonstrated that Sri Lankan elephants are attracted to the agricultural landscape as a food source. They have the ability to recognize that seasonal croplands offer a substantial food supply for wild elephants. In the dry zone of Sri Lanka, where the majority of elephants reside, rice paddy farming is a prevalent practice. The specific farming practices employed depend on the availability of seasonal water.

Dry-zone agriculture in Sri Lanka encompasses various farming practices, including intensive rice-rice cultivation, rice-vegetable farming, or cultivation of rice along with other field crops such as mungbean, millets, maize, chili, soybean, and seasonal vegetables. Additionally, rice-fallow rotations are also observed in the region [51]. Therefore, within the dry zone, the Sri Lankan elephant benefits from a combination of habitats, including the dry evergreen forests, agricultural lands, and croplands. Elephants utilize the forested areas for cover and browsing opportunities, while farming areas provide accessible and abundant grazing opportunities [25, 33, 47, 52].

Previous research and observations have highlighted the importance of these diverse habitats for the Sri Lankan elephant population. The elephants rely on the mixture of forested and agricultural landscapes, utilizing each habitat for specific purposes. The forested areas offer protective cover and browsing resources, while the agricultural and cropland habitats provide ample grazing opportunities. This interplay between dry evergreen forests and farming areas contributes to the elephants' overall foraging strategies and supports their sustenance in the dry zone.

4. Crop-raiding behavior of elephants

Crop-raiding behavior is a common occurrence among elephants, but it is often misunderstood that elephants resort to raiding cultivated crops due to a lack of natural food sources. In reality, elephants maintain large home ranges that often overlap with human-dominated landscapes, leading them to feed on human-cultivated crops based on seasonality and availability [53]. However, this exposes elephants to various threats from people, particularly in agricultural-dominated landscapes. Consequently, elephants have adapted their behaviors and activities to reduce risks and maximize benefits in human landscapes.

A significant adaptation by elephants to coexist with humans is their transition to diurnal behavior. Elephants are predominantly active from late afternoon to early morning and seek refuge away from human-dominated areas during the daytime. They venture into human landscapes during the cover of darkness and return to forested areas in the morning [13, 25]. Elephants primarily move toward open natural habitats such as grass beds near man-made lakes and scrublands to feed. They approach crop fields only during complete darkness [13, 25, 54, 55], which poses a considerable risk for humans.

Several studies have suggested that feeding on cultivated crops provides elephants with better nutrients compared to foraging on natural food sources [13, 56, 57]. However, the scientific evidence supporting this claim is not yet comprehensive. Nevertheless, Fernando et al. [25] clearly demonstrated that elephant incursions into human landscapes were more frequent during the harvesting periods of rice and other crops, particularly watermelon, banana, and corn. This indicates that elephants are attracted to these crops due to their nutritional value (**Figure 3**). It is important to note that natural fodder for elephants is not scarce during these times. Farmers intensively guard their crops, exposing elephants to high threats. Despite the risks involved, elephants' fondness for crop fields suggests the potential benefits in terms of acquiring nutrients, albeit at the expense of increased interaction with humans [25, 58].

Risk-taking behavior varies among different social groups of elephants [25, 53, 55, 59]. Family groups tend to minimize risks as they are accompanied by juveniles of different ages. For example, we observed a scarcity of mother-calf units entering crop fields and feeding on crops. Consequently, they are less frequently found in crop fields, choosing to sacrifice the gains from cultivated crops to protect their young. Our observations of two-family groups equipped with GPS collars in Southern Sri Lanka revealed that they stayed in small forest patches among crop fields and abandoned agricultural fields more often than entering the crop fields. Additionally, our observations from different locations in Sri Lanka showed that family groups would enter paddy fields immediately after harvest to consume leftover paddy, as these fields are unguarded by farmers, resulting in no human-related risks.

When it comes to the trade-off decision of taking high risks to reap the benefits of foraging on cultivated crops, male elephants are known to be more inclined toward risk-taking behavior. Sexual segregation in movement patterns toward human landscapes, based on crop availability, indicates that males are more willing to take risks [25, 60]. It has been suggested that the high-risk, high-gain foraging behavior of male elephants contributes to their growth and better body condition, ultimately enhancing their reproductive capacity by obtaining cultivated food items [57, 61]. Another possible reason for males to feed on cultivated crops more than females is their larger home ranges, which necessitate more extensive use of human-dominated landscapes, including crop fields while seeking natural resources within their home ranges [62, 63]. Furthermore, according to [53], elephants raid crops not as a foraging strategy but because their natural foraging resources in their home ranges have been depleted due to human activities, leaving them with limited feeding options. This idea is supported by research findings on the ranging patterns of elephants by Fernando [64], which indicate that elephants in Sri Lanka do not engage in long-distance migrations or maintain wet- and dry-season ranges. Consequently, they must find resources within their home ranges. Both scenarios of feeding on crops are equally valid based on our observations.

It is crucial to note that not all elephants engage in crop raiding. However, in rural Sri Lanka, there exists a general perception among farmers that all elephants

Figure 3.
Crop raid by elephants. (A) Crop damage after elephant raid in Sri Lanka, (B) Elephant dung in a crop field after raiding the crop overnight.

raid crops and that all male elephants are crop raiders, considering them a nuisance and serious pests [24]. Nevertheless, this is not the case, as not all elephants are crop raiders; only certain individuals and family groups engage in crop raiding. Desai and Heidi [53] classified crop-raiding elephants into three categories: opportunistic crop raiders, habitual crop raiders, and obligatory crop raiders. Opportunistic crop raiders are elephants that raid crops when they come across unprotected fields. Most of these opportunistic raiders are family groups. Habitual crop raiders are individuals that develop a liking for crops and become regular raiders, particularly young and mature male elephants. These elephants learn to overcome human-elephant conflict (HEC) mitigation measures implemented by farmers. Male habitual raiders form small bachelor groups when raiding crops, which may serve as a behavioral strategy to collectively face human threats. Young male elephants join bachelor groups with mature males to learn from them. Obligatory crop raiders are elephants that have experienced significant losses of their natural habitats and foraging resources due to habitat clearance, thus being forced to depend to a greater extent on cultivated crops. These elephants frequently encounter humans and are more vulnerable to human retaliations.

Currently, elephants face threats solely due to their status as megaherbivores living within their home ranges. They are compelled to exist in highly fragmented habitats embedded in human-dominated landscapes, leading to increased human actions to control their entry. In Sri Lanka, farmers take invasive measures more frequently than in the past to control elephants that encroach upon and linger around their crop fields.

5. Elephants eating garbage

Human activities have exerted significant pressure on elephants, leading to extreme changes in their food preferences. In Sri Lanka, some elephants have shifted their feeding habits from foraging on natural food sources to consuming food items found in garbage sites. These garbage dumps offer a plentiful supply of both organic and inorganic waste. Organic waste includes discarded fruits, vegetables, and cooked food, while inorganic materials consist of plastics, glass, metals, and other synthetic items. Elephants are primarily attracted to easily accessible organic matter but may accidentally consume inorganic substances.

Our observations of elephants at various garbage sites indicated that only mature and sub-adult males were present at the dumps, while female groups with juveniles did not visit these locations (**Figure 4**). The only published study on garbage site observations in Bundala, Southern Sri Lanka, conducted by Liyanage et al. [65], supports these findings. Once elephants become accustomed to feeding at these sites, they establish a permanent presence in the vicinity of the garbage dumps. Consequently, their reliance on natural food sources diminishes, and their home range becomes significantly reduced compared to other elephants. These elephants obtain sufficient food quantity and essential nutrients from the items found in the garbage dump, which is reflected in their healthy body condition score. There is a common perception that these elephants should be deterred from consuming garbage due to potential harm to their well-being and the unpleasant sight of elephants in such environments. However, conservationists argue that if these elephants are prevented from accessing garbage sites, they will resort to frequent crop raiding, exposing them to life-threatening injuries and potential retaliatory actions by humans. Therefore, it is preferable to allow these elephants to feed at the garbage sites, with authorities taking measures to separate them from inorganic materials before disposal to minimize accidental ingestion.

In addition to feeding on garbage, elephants in Sri Lanka have adopted another foraging strategy, which involves soliciting food from people along roadsides. These elephants are known as "road elephants" (**Figure 5**). They can often be observed along the Buttala-Kataragama road in Southern Sri Lanka. The elephants select

DOI: http://dx.doi.org/10.5772/intechopen.112758

Figure 4.
A wild elephant waiting for hands outs of food from passing traffic, blocks the Kataragama road Buttala Sri Lanka.

Figure 5.
Sri Lankan wild elephant herd eating garbage at Wellawaya, Sri Lanka.

specific spots near the road and wait for passing traffic to request food. Initially, this behavior was primarily observed in a few mature males, but recently, in 2019, we observed a family group, including females and juveniles, imitating the males and engaging in begging behavior only during the late afternoon due to their fear of humans and traffic. Several family groups have permanently stationed themselves by the road with juveniles, raising significant conservation concerns. The begging behavior exhibited by male elephants is learned by other elephants and transmitted to different populations.

These unique feeding strategies, involving garbage consumption and begging from people, highlight the adaptive behavior of elephants in response to human-altered landscapes. Similar patterns of garbage feeding and road begging have been observed in other parts of the world, indicating the widespread nature of these behavioral adaptations These observations emphasize the urgent need for comprehensive conservation measures to address the underlying causes of these behavioral changes, mitigate human-elephant conflicts, and develop effective strategies for protected area management and conservation biology of elephants. Understanding and managing these adaptive behaviors is crucial for the long-term survival and coexistence of elephants and humans in shared landscapes.

6. Impact on habitats

Larger herbivores play a crucial role as ecosystem engineers, modifying habitats through their feeding behaviors and exerting scale-dependent effects on natural environments [66–68]. Elephants, as one of the largest herbivores, have significant impacts on habitat composition, particularly on woody plants, by breaking branches, toppling trees, and reducing the recruitment of new vegetation [69]. These modifications are essential drivers of habitat dynamics and can lead to the generation of new habitats that support ecological restoration.

The impact of elephants on woody plants is influenced by various factors, including overall food availability, food preferences, spatial and temporal weather conditions, social interactions, and the physical status of elephants [70–72]. For instance, in Gonarezhou National Park in Zimbabwe, the browsing behavior of elephants resulted in the degradation of woodland patches dominated by *Acacia tortillis* [73]. Research indicated that bull elephants, particularly those of higher density, were responsible for more significant damage to trees compared to family groups [74]. High population densities of elephants can exert severe pressure on the survival of threatened plants and disrupt plant community structures, as observed in Addo National Park in South Africa [75]. Efforts are being made in Greater Kruger National Park to protect keystone tree species like marula (*Sclerocarya birrea*) from elephant impact.

In Asia, studies by Sukumar [76] demonstrated the browsing effects of elephants on woody plants, revealing variations in foraging impact across different tree size classes. Unlike in Africa, the specific aspects of elephant foraging ecology and their habitat impacts have received less attention in Asian contexts [25, 53, 60]. In Sri Lanka, we have observed that the elephants have not caused significant habitat changes through their foraging behaviors, as seen in Africa. However, they do exhibit selectivity in targeting certain tree species. In Wasgamuwa, for example, elephants selectively forage on two woody plant species: Maila (*Bauhinia racemose*) and Welang tree (*Pterospermum suberifolium*). Elephants debark Maila trees after monsoon rains, with family groups contributing to much of the damage. Additionally, Maila trees are selectively targeted during the dry season when mature seed pods are abundant, with elephants breaking branches to acquire the sweet seeds. The impact on Maila trees does not have a negative effect on the habitat; instead, the thinning of Maila trees helps maintain the scrubby nature of the habitat, preventing or delaying the succession from scrubland to forest.

These examples illustrate the significant role of elephants as ecological influencers, shaping habitats through their feeding behaviors and highlighting the dynamic relationship between elephants and their environments. These relationships are of the utmost importance for protected area management and conservation biology. Understanding the ecological effects of elephants' foraging behaviors helps inform conservation strategies aimed at maintaining biodiversity and ecosystem functioning within protected areas. By altering habitat composition through branch breaking, tree toppling, and selective feeding, elephants shape the structure and dynamics of plant communities, which in turn influences the availability of resources for other species. These habitat modifications can have cascading effects on other wildlife, including changes in species composition, plant regeneration, and overall ecosystem resilience.

Therefore, incorporating the knowledge of elephant impacts into protected area management plans allows for the implementation of targeted conservation measures to mitigate potential negative effects, ensure the survival of threatened plant species, and maintain the integrity of habitats. Additionally, understanding the role of elephants as ecosystem engineers and the generation of new habitats through their behaviors provides valuable insights for ecological restoration efforts, guiding the restoration of degraded areas and promoting the establishment of diverse and functional ecosystems. By recognizing and addressing the complex interactions between elephants, their habitats, and other species, protected area managers and conservation biologists can effectively conserve biodiversity and promote the sustainable coexistence of elephants and humans.

7. Emerging threats to elephants

The high-risk foraging strategy of male elephants is identified as the primary cause of human-elephant conflict (HEC) [57, 62, 64]. In Sri Lanka, the majority of elephant deaths occur in landscapes dominated by agriculture [77]. While various measures have been employed to mitigate elephant crop damage, there has been an alarming increase in the utilization of highly destructive methods to control elephants, posing a significant concern for elephant conservation [77]. We aim to provide details on some of these destructive measures to shed light on this pressing issue.

One of the most destructive methods used to control elephants is the deployment of "jaw bombs." These homemade explosive devices, resembling small grenades, are created by combining gunpowder with tightly packed stones or metal pieces. Initially developed to combat agricultural pests like wild boar, jaw bombs are designed to detonate within an animal's mouth when inadvertently bitten. The explosion causes severe damage to the jaws and tongue. Disturbingly, villagers have started hiding jaw bombs inside fruits, such as watermelons, to target elephants. Often, these devices are strategically placed at the periphery of crop fields to deter raiders. It is worth noting that the majority of jaw bomb victims are female and young elephants. This observation suggests that male elephants may have developed a heightened sense to detect and avoid these concealed explosives. However, further scientific investigation is required to confirm this hypothesis.

Sadly, most elephants affected by jaw bombs are unable to eat or drink due to their injuries, leading to minimal chances of survival from a veterinary standpoint. Troublingly, the use of jaw bombs is becoming increasingly prevalent and widespread in Sri Lanka. Farmers and individuals deploy these explosives in nearly all crop fields, predominantly during the night and retrieving them in the morning. Given that elephants are active during the night, they are particularly vulnerable to encountering jaw bombs. Consequently, we strongly recommend that government authorities take decisive action to prevent and combat this illegal activity.

The second highly destructive method employed to control crop-raiding elephants is electrocution. Farmers and villagers' resort to this method to safeguard their crops and properties. Typically, this involves setting up single metal strings around the targeted fields, connected to electrical power sources from nearby houses. The continuous high amperage running through the metal strings poses a fatal threat to elephants and any other animals, including humans, coming into contact with them. Electric shocks lead to ventricular fibrillation, causing heart failure, burns, and internal organ damage. It is crucial to emphasize that electrocution is an illegal practice. Recognizing its illegality, villagers predominantly set up these lethal traps at night when elephants are active, removing them during the daytime. Male elephants constitute the majority of victims of electrocution. To mitigate the use of such methods, farmers should be encouraged to participate in crop-guarding initiatives, including the establishment of village and crop-field fences. It is disheartening to note that some farmers resort to additional invasive measures, such as shooting elephants with homemade illegal guns, employing gun traps, and setting up metal snares, to harm or kill elephants in agricultural landscapes.

Addressing the issue of destructive measures used to control elephants is paramount within the framework of protected area management and conservation biology in Sri Lanka. It requires a comprehensive approach that encompasses several key aspects. Firstly, there is a pressing need for increased law enforcement and strict regulation to curb illegal activities related to jaw bombs, electrocution, and other harmful methods. Government authorities must take decisive action to halt the production, distribution, and use of these destructive devices, ensuring that those involved in such practices face legal consequences. This includes strengthening patrols and surveillance in vulnerable areas, collaborating with local communities to gather information and intelligence, and establishing mechanisms to encourage individuals to report any instances of illegal elephant control methods.

The island's elephant populations face potential threats from current and future irrigation projects. One such example is the completion of the Moragahakanda Dam, which has had detrimental effects on the elephant population in Minneriya. Unseasonal water releases during the dry season have disrupted the natural water levels in the reservoir, leading to a decline in elephants. Despite the presence of limited grasslands, even a minor increase in water levels prompts several elephant herds to move away from Minneriya. As a result, there has been a significant decrease in the number of elephants observed during the peak of the "elephant gathering" over the years. It is crucial to closely monitor these populations and study their feeding behavior and movement patterns to understand the full impact of these changes.

To safeguard the long-term conservation future of the elephant population and preserve the natural wonders of Minneriya, it is imperative to address the issues arising from the Moragahakanda Dam project. Additionally, it is essential to ensure that future development projects undergo thorough environmental assessments and planning processes. By taking these measures, we can mitigate the adverse effects on elephants and maintain the ecological balance of this unique habitat.

8. Conclusions and future directions

The historical and ongoing interaction between elephants and humans in Sri Lanka has resulted in significant challenges for elephant conservation. Development projects in the country's dry zone have led to the loss and fragmentation of elephant habitats, pushing them into more marginal areas. As a response, elephants have exhibited both positive and negative behavioral adaptations to survive in human-dominated landscapes, particularly through their feeding habits. However, these adaptations have also contributed to the escalating issue of human-elephant conflict (HEC), which remains the leading cause of daily elephant fatalities in Sri Lanka. It is crucial to prioritize research efforts aimed at enhancing favorable habitats within elephants' home ranges, preserving natural wildlife corridors, and implementing sustainable and effective crop protection measures to mitigate crop raiding and conserve Sri Lankan elephants.

Education and awareness programs play a pivotal role in fostering coexistence and changing societal attitudes and behaviors toward elephants. Public outreach initiatives should focus on highlighting the negative impacts of destructive measures, raising awareness about the ecological importance of elephants, and promoting alternative conflict mitigation strategies that are both humane and sustainable. Engaging local communities, farmers, and landowners through capacity-building programs and providing support for the implementation of effective crop protection measures, such as physical barriers, deterrents, and early warning systems, can significantly reduce reliance on harmful methods while safeguarding livelihoods and ensuring food security.

Additionally, scientific research and monitoring are essential for understanding the complex dynamics between elephants, their habitats, and human activities. Investigating the ecological impacts of elephant foraging behaviors, movement patterns, and responses to various deterrent methods can inform evidence-based management strategies. Close collaboration among conservation biologists, researchers, protected area managers, and government agencies is crucial for generating scientific knowledge that guides conservation efforts and supports the development of sustainable policies and practices.

By integrating these approaches into protected area management and conservation biology, Sri Lanka can make significant progress in mitigating human-elephant conflict, protecting elephants from harmful measures, and promoting a harmonious coexistence between humans and these majestic creatures. This holistic approach not only ensures the survival of elephants but also contributes to the conservation of biodiversity, the maintenance of ecosystem functioning, and the preservation of Sri Lanka's rich natural heritage for future generations to enjoy.

Acknowledgements

We are grateful to all the villages who shared their knowledge and experience of elephants with us. We would like to thank the Sri Lankan Wildlife Conservation Society (SLWCS) staff and the Centre for Conservation and Research (CCR). And a special thank goes to the Department of Wildlife Conservation (DWC) staff in Wasgamuwa. The corresponding author would like to thanks to the Institute for Applied Ecology, University of Canberra, for generously covering the Article Processing Charge for this publication.

Conflict of interest

The author declares no conflict of interest.

Author details

Chandima Fernando[1,2] and Duminda S.B. Dissanayake[3*]

1 Faculty of Science, Department of Zoology and Environmental Science, University of Colombo, Colombo, Sri Lanka

2 Sri Lanka Wildlife Conservation Society, Nugegoda, Sri Lanka

3 Institute for Applied Ecology, University of Canberra, Canberra, Australia

*Address all correspondence to: dumie.dissanayake@canberra.edu.au

References

[1] Cumming DHM, Cumming GS. Ungulate community structure and ecological processes: Body size, hoof area and trampling in African savannas. Oecologia. 2003;**134**(4):560-568

[2] Malhi Y, Doughty CE, Galetti M, Smith FA, Svenning JC, Terborgh JW. Megafauna and ecosystem function from the Pleistocene to the Anthropocene. Proceedings of the National Academy of Sciences. 2016;**113**(4):838-846

[3] Heggenes J, Odland A, Chevalier T, Ahlberg J, Berg A, Larsson H, et al. Herbivore grazing—Or trampling? Trampling effects by a large ungulate in cold high-latitude ecosystems. Ecology and Evolution. 2017;**7**(16):6423-6431

[4] WallisDeVries MF. Large herbivores as key factors for nature conservation. In: WallisDeVries MF, Van Wieren SE, Bakker JP, editors. Grazing and Conservation Management [Internet]. Dordrecht: Springer Netherlands; 1998. p. 1-20 [cited 2022 Sep 29]. (Conservation Biology Series). DOI: 10.1007/978-94-011-4391-2_1

[5] Ripple WJ, Newsome TM, Wolf C, Dirzo R, Everatt KT, Galetti M, et al. Collapse of the world's largest herbivores. Science Advances. 2015;**1**(4):e1400103

[6] Ahrestani FS, Sankaran M. Introduction: The large herbivores of south and Southeast Asia—A prominent but neglected guild. In: Ahrestani FS, Sankaran M, editors. The Ecology of Large Herbivores in South and Southeast Asia [Internet]. Dordrecht: Springer Netherlands; 2016. pp. 1-13. [cited 2022 Sep 29] (Ecological Studies). DOI: 10.1007/978-94-017-7570-0_1

[7] Ahrestani FS, Heitkönig IMA, Matsubayashi H, Prins HHT. Grazing and browsing by large herbivores in south and Southeast Asia. In: Ahrestani FS, Sankaran M, editors. The Ecology of Large Herbivores in South and Southeast Asia [Internet]. Dordrecht: Springer Netherlands; 2016. pp. 99-120. [cited 2022 Sep 29] (Ecological Studies). DOI: 10.1007/978-94-017-7570-0_4

[8] Sukumar R. A brief review of the status, distribution and biology of wild Asian elephants Elephas maximus. International Zoo Yearbook. 2006;**40**(1):1-8

[9] Kemf E, Santiapillai C. Asian Elephants in the Wild. A WWF Species Status Report. Gland, Switzerland: WWF International; 2000

[10] Menon V, Tiwari SKr. Population status of Asian elephants Elephas maximus and key threats. International Zoo Yearbook. 2019;**53**(1):17-30

[11] IUCN. The IUCN red list of threatened species [Internet]. 2022 [cited 2022 Oct 6]. Available from: https://www.iucnredlist.org/

[12] Olivier R. Distribution and status of the Asian elephant. Oryx. 1978;**14**(4):379-424

[13] Sukumar R. The Asian Elephant: Ecology and Management. Cambridge, UK: Cambridge University Press; 1989. p. 276

[14] Shoshani J, Eisenberg JF. Elephas maximus. Mammalian Species. 1982;**182**:1-8

[15] Leimgruber P, Gagnon JB, Wemmer C, Kelly DS, Songer MA,

Selig ER. Fragmentation of Asia's remaining wildlands: Implications for Asian elephant conservation. Animal Conservation. 2003;**6**(4):347-359

[16] McKay GM. Behavior and ecology of the Asiatic elephant in southeastern Ceylon. 1973 [cited 2022 Oct 4]; Available from: http://repository.si.edu/xmlui/handle/10088/5383

[17] Ministry of Agriculture. Wild Elephant Population in Sri Lanka Swell up to 7000. Colombo, Sri Lanka: Ministry of Agriculture and Department of Wildlife Conservation; 2022. [Internet] Available from: https://www.agrimin.gov.lk/web/index.php/news-scroll/1804-2022-09-15-1e?lang=en

[18] LaDue CA, Vandercone RPG, Kiso WK, Freeman EW. Social behavior and group formation in male Asian elephants (Elephas maximus): The effects of age and Musth in wild and zoo-housed animals. Animals. 2022;**12**(9):1215

[19] Santiapillai C, Wijeyamohan S, Bandara G, Athurupana R, Dissanayake N, Read B. An assessment of the human-elephant conflict in Sri Lanka. Ceylon Journal of Science Biological Sciences. 2010;**39**(1):21-33

[20] Fernando P, Jayewardene J, Prasad T, Hendavitharana W, Pastorini J. Current status of Asian Elephants in Sri Lanka. 2011 [cited 2022 Oct 4] Available from: https://www.zora.uzh.ch/id/eprint/59037

[21] Köpke S, Withanachchi SS, Pathiranage R, Withanachchi CR, Gamage DU, Nissanka TS, et al. Human-elephant conflict in Sri Lanka: A critical review of causal explanations. Sustainability. 2021;**13**(15):8625

[22] Prakash TGSL, Wijeratne AW, Fernando P. Human-elephant conflict in Sri Lanka: Patterns and extent. Gajah. 2020;**51**:16-25

[23] Perera B. The human-elephant conflict: A review of current status and mitigation methods. Gajah. 2009;**30**:41-52

[24] Fernando P, Wikramanayake E, Weerakoon D, Jayasinghe LKA, Gunawardene M, Janaka HK. Perceptions and patterns of human–elephant conflict in old and new settlements in Sri Lanka: Insights for mitigation and management. Biodiversity and Conservation. 2005;**14**(10):2465-2481

[25] Fernando C, Weston MA, Corea R, Pahirana K, Rendall AR. Asian elephant movements between natural and human-dominated landscapes mirror patterns of crop damage in Sri Lanka. Oryx. 2022:**4**:1-8

[26] Rathnayake CWM, Jones S, Soto-Berelov M, Wallace L. Human–elephant conflict and land cover change in Sri Lanka. Applied Geography. 2022;**143**:102685

[27] De Silva MDS. Status and conservation of the elephant and the alleviation of man-elephant conflict in Sri Lanka. Gajah. 1998;**19**:1-24

[28] Jayawardene J. Elephant and Mahaweli: A 15-year study. Sri Lanka Nature. 1998;**2**:45-51

[29] Bandara R, Tisdell C. Asian elephants as agricultural pests: Economics of control and compensation in Sri Lanka. Natural Resources Journal. 2002;**42**(3):491-519

[30] Survey Department of Sri Lanka, editor. The National Atlas of Sri Lanka. 2nd ed. Colombo, Sri Lanka: Survey Department of Sri Lanka; 2012. p. 234

[31] Ishwaran N. Ecology of the Asian elephant in lowland dry zone habitats of the Mahaweli River basin, Sri Lanka. Journal of Tropical Ecology. 1993;**9**(2):169-182

[32] Pastorini J, Janaka HK, Nishantha HG, Prasad T, Leimgruber P, Fernando P. A preliminary study on the impact of changing shifting cultivation practices on dry season forage for Asian elephants in Sri Lanka. Tropical Conservation Science. 2013;**6**(6):770-780

[33] Samansiri KAP, Weerakoon DK. Feeding behaviour of Asian elephants in the Northwestern region of Sri Lanka. Gajah. 2007;**27**:27-348

[34] Weerakoon D, Gunawardene M, Janaka H, Jayasinghe L, Perera R, Fenando P, et al. Ranging BehaviourAnd habitat use of elephants in Sri Lanka. In: Proceedings of the Symposium for Human-Elephant Relationships and Conflicts. Colombo: Biodiversity and Elephant Conservation Trust; 2003. pp. 68-70

[35] Kumar MA, Mudappa D, Raman TRS. Asian elephant Elephas Maximus habitat use and ranging in fragmented rainforest and plantations in the Anamalai Hills, India. Tropical Conservation Science. 2010;**3**(2):143-158

[36] Blake S, Maisels F. Forest elephant movements in Central Africa: Megafauna need Megaspaces. In: Reyna-Hurtado R, Chapman CA, Melletti M, editors. Movement Ecology of Afrotropical Forest Mammals. [Internet]. Cham: Springer International Publishing; 2023. pp. 27-58. [cited 2023 Jun 26]. DOI: 10.1007/978-3-031-27030-7_3

[37] Laguardia A, Gobush KS, Bourgeois S, Strindberg S, Abitsi G, Ebouta F, et al. Assessing the feasibility of density estimation methodologies for African forest elephant at large spatial scales. Global Ecology and Conservation. 2021;**27**:e01550

[38] Campos-Arceiz A, Blake S. Megagardeners of the forest – The role of elephants in seed dispersal. Acta Oecologica. 2011;**37**(6):542-553

[39] Tan WH, Hii A, Solana-Mena A, Wong EP, Loke VPW, Tan ASL, et al. Long-term monitoring of seed dispersal by Asian elephants in a Sundaland rainforest. Biotropica. 2021;**53**(2):453-465

[40] Mahesha P, Rajnish V. Some aspects of seed dispersal by Asian elephants (Elephas maximus) in Kaudulla National Park, Sri Lanka. Current Science. 2020;**118**(4):648-654

[41] Samansiri KAP, Weerakoon DK. A study on the seed dispersal capability of Asian elephants in the Northwestern region of Sri Lanka. Gajah. 2008;**28**:19-24

[42] Chathuranga WGD, Ranawana KB. A preliminary investigation of seed dispersal by elephants (Elephas maximus maximus) in Kumaragala Forest reserve, Matale District, Sri Lanka. Ceylon Journal of Science. 2017;**46**(3):39-46

[43] Haynes G. Elephants (and extinct relatives) as earth-movers and ecosystem engineers. Geomorphology. 2012;**157-158**:99-107

[44] Erdelen W. Forest ecosystems and nature conservation in Sri Lanka. Biological Conservation. 1988;**43**(2):115-135

[45] Gunawardana RALH. Irrigation and hydraulic society in early medieval Ceylon. Past Present. 1971;**53**:3-27

[46] Pastorini J, Nishantha HG, Janaka HK, Isler K,

Fernando P. Water-body use by Asian elephants in southern Sri Lanka. Tropical Conservation Science. 2010;**3**(4):412-422

[47] Alahakoon AMDB, Pushpakumara EMAB, Ellepola G, Ranawana KB. Food and feeding patterns of Asian elephants in Udawalawe National Park, Sri Lanka. Gajah. 2017;**46**:4-13

[48] CodronJ,Lee-ThorpJA,SponheimerM, Codron D, Grant RC, de Ruiter DJ. Elephant (Loxodonta africana) diets in Kruger National Park, South Africa: Spatial and landscape differences. Journal of Mammalogy. 2006;**87**(1):27-34

[49] Mueller-Dombois D. Crown distortion and elephant distribution in the Woody Vegetations of Ruhuna National Park, Ceylon. Ecology. 1972;**53**(2):208-226

[50] Hewavisenthi ACDS. Mahaweli Water resources project. Water International. 1992;**17**(1):33-43

[51] Ratnayake RR, Perera BMACA, Rajapaksha RPSK, Ekanayake EMHGS, Kumara RKGK, Gunaratne HMAC. Soil carbon sequestration and nutrient status of tropical rice based cropping systems: Rice-rice, rice-soya, rice-onion and rice-tobacco in Sri Lanka. Catena. 2017;**150**:17-23

[52] Ishwaran N. Elephant and woody-plant relationships in gal Oya, Sri Lanka. Biological Conservation. 1983;**26**(3):255-270

[53] Desai A, Riddle H. Human-Elephant Conflict in Asia. [Internet]. Washington, DC, USA: U.S. Fish and Wildlife Service; 2015. Available from: https://pdfs.semanticscholar.org/2259/a8cce2754037bc44aac29ce283e5f9f38aeb.pdf

[54] Sitati NW, Walpole MJ, Smith RJ, Leader-Williams N. Predicting spatial aspects of human–elephant conflict. Journal of Applied Ecology. 2003;**40**(4):667-677

[55] Graham MD, Douglas-Hamilton I, Adams WM, Lee PC. The movement of African elephants in a human-dominated land-use mosaic. Animal Conservation. 2009;**12**(5):445-455

[56] Barnes RFW, Dubiure UF, Danquah E, Boafo Y, Nandjui A, Hema EM, et al. Crop-raiding elephants and the moon. African Journal of Ecology. 2007;**45**(1):112-115

[57] Chiyo PI, Lee PC, Moss CJ, Archie EA, Hollister-Smith JA, Alberts SC. No risk, no gain: Effects of crop raiding and genetic diversity on body size in male elephants. Behavioral Ecology. 2011;**22**(3):552-558

[58] Wilson S, Davies TE, Hazarika N, Zimmermann A. Understanding spatial and temporal patterns of human–elephant conflict in Assam, India. Oryx. 2015;**49**(1):140-149

[59] Srinivasaiah N, Kumar V, Vaidyanathan S, Sukumar R, Sinha A. All-male groups in Asian elephants: A novel, adaptive social strategy in increasingly anthropogenic landscapes of southern India. Scientific Reports. 2019;**9**(1):8678

[60] Vogel SM, Lambert B, Songhurst AC, McCulloch GP, Stronza AL, Coulson T. Exploring movement decisions: Can Bayesian movement-state models explain crop consumption behaviour in elephants (Loxodonta africana)? The Journal of Animal Ecology. 2020;**89**(4):1055-1068

[61] Sukumar R, Gadgil M. Male-female differences in foraging on crops by Asian elephants. Animal Behaviour. 1988;**36**(4):1233-1235

[62] Jackson TP, Mosojane S, Ferreira SM, van Aarde RJ. Solutions for elephant

Loxodonta africana crop raiding in northern Botswana: Moving away from symptomatic approaches. Oryx. 2008;**42**(1):83-91

[63] Songhurst A, Coulson T. Exploring the effects of spatial autocorrelation when identifying key drivers of wildlife crop-raiding. Ecology and Evolution. 2014;**4**(5):582-593

[64] Fernando P. Elephants in Sri Lanka: Past present and future. Loris. 2000;**22**(2):38-44

[65] Liyanage DJ, Fernando P, Dayawansa PN, Janaka HK, Pastorini J. The elephant at the dump: How does garbage consumption impact Asian elephants? Mammalian Biology. 2021;**101**(6):1089-1097

[66] Jones CG, Lawton JH, Shachak M. Organisms as ecosystem engineers. Oikos. 1994;**69**(3):373-386

[67] Pringle RM. Elephants as agents of habitat creation for small vertebrates at the patch scale. Ecology. 2008;**89**(1):26-33

[68] Coverdale TC, Kartzinel TR, Grabowski KL, Shriver RK, Hassan AA, Goheen JR, et al. Elephants in the understory: Opposing direct and indirect effects of consumption and ecosystem engineering by megaherbivores. Ecology. 2016;**97**(11):3219-3230

[69] Western D, Maitumo D. Woodland loss and restoration in a savanna park: A 20-year experiment. African Journal of Ecology. 2004;**42**(2):111-121

[70] Sach F, Dierenfeld ES, Langley-Evans SC, Watts MJ, Yon L. African savanna elephants (Loxodonta africana) as an example of a herbivore making movement choices based on nutritional needs. PeerJ. 2019;**7**:e6260

[71] de Beer Y, Kilian W, Versfeld W, van Aarde RJ. Elephants and low rainfall alter woody vegetation in Etosha National Park, Namibia. Journal of Arid Environments. 2006;**64**(3):412-421

[72] van de Water A, Henley M, Bates L, Slotow R. The value of elephants: A pluralist approach. Ecosystem Services. 2022;**58**:101488

[73] Gandiwa E, Magwati T, Zisadza P, Chinuwo T, Tafangenyasha C. The impact of African elephants on Acacia tortilis woodland in northern Gonarezhou National Park, Zimbabwe. Journal of Arid Environments. 2011;**75**(9):809-814

[74] Hiscocks K. The impact of an increasing elephant population on the woody vegetation in southern Sabi sand Wildtuin, South Africa. Koedoe. 1999;**42**(2):47-55

[75] Lombard AT, Johnson CF, Cowling RM, Pressey RL. Protecting plants from elephants: Botanical reserve scenarios within the Addo elephant National Park, South Africa. Biological Conservation. 2001;**102**(2):191-203

[76] Sukumar R. Ecology of the Asian elephant in southern India. II. Feeding habits and crop raiding patterns. Journal of Tropical Ecology. 1990;**6**(1):33-53

[77] De Silva MDS. Spatial patterns and major causes of elephant mortalities in the Mahaweli wildlife conservation area, SriLanka [M.Sc]. Kandy, Sri Lanka: University of Peradeniya; 2020